Mechanism of Composite Explosion Inhibition and Venting Protections

Zhirong Wang · Xingyan Cao

Mechanism of Composite Explosion Inhibition and Venting Protections

Springer

Zhirong Wang
College of Emergency Management
Nanjing Tech University
Nanjing, China

Xingyan Cao
College of Safety Science and Engineering
Nanjing Tech University
Nanjing, China

ISBN 978-981-95-4808-8 ISBN 978-981-95-4809-5 (eBook)
https://doi.org/10.1007/978-981-95-4809-5

This work was supported by National Natural Science Foundation of China (52274210; 51806056; 52074159) and The Open Fund of the State Key Laboratory of Fire Science (SKLFS) Program (HZ2025-KF09). The authors are grateful for National project funding for Key R & D programs under Grant No. 2023YFC3009900, and Prof. Zhirong Wang and Prof. Xingyang Cao are Research Fellows at the Institute of Emergency Governance and Policy, Nanjing Tech University, China.

This Springer imprint is published by the registered company Springer Nature Singapore Pte Ltd.
The registered company address is: 152 Beach Road, #21-01/04 Gateway East, Singapore 189721, Singapore

The original version of the book has been revised. A correction to this book can be found at https://doi.org/10.1007/978-981-95-4809-5_10

Competing Interests The authors have no competing interests to declare that are relevant to the content of this manuscript.

Contents

Chapter 1
Introduction

Explosion accident occurs frequently and explosion form becomes more complex with the rapid developments of industry and society. And the explosion form gradually changes from a single component to an explosion composed of combustible substances with multiple components, especially the gas and dust hybrid explosions [1, 2]. The process of gas/dust hybrid explosions is more complex and the intensity of explosion is also greater compared with the single component, further resulting the more serious life and property losses [3–5]. Hence, realizing gas/dust hybrid explosions inhibition and reducing the economic losses and casualties caused by the hybrid explosions has become the key to be solved urgently. The explosion inhibiting and venting are two effective protection methods and is widely used in the petrochemical industry.

The type and amount of inhibitor and the determination of inhibition mechanism need to consider the applicability of combustible substance, inhibition effect and process environment. Considering the characteristics of gas/dust hybrid explosions, the water mist, especially the ultrafine water mist that has a larger specific surface area (i.e., droplet diameter below 50 μm) and good application prospects due to its high efficient endothermic properties, has been considered as an ideal explosion inhibitor. Scholars have carried out the relevant experimental research on the inhibitions of methane and coal dust hybrid explosions. The inhibition of gas/dust hybrid explosions had also been studied accordingly. Gan et al. [6] pointed out that ultrafine water mist could reduce the flame propagation velocity of PMMA dust explosion, and the flame temperature and its rising rate were also obviously decreased. Feng et al. [7] studied the inhibition effect of ultrafine water mist on methane and coal dust hybrid explosions and found that the pressure of gas/dust hybrid explosions was decreased obviously. Meanwhile, the contribution of fixed carbon combustion to explosion pressure was reduced by 50.8% under lean burn condition, and the contribution of volatile matter to explosion pressure was reduced by 36.3% and 40.8% respectively under the equivalence ratio and rich burn conditions. Chen et al. [8] found that the time for the hybrid explosions pressure history began to rise was prolonged.

Z. Wang and X. Cao, *Mechanism of Composite Explosion Inhibition and Venting Protections*, https://doi.org/10.1007/978-981-95-4809-5_1

And the flame propagation velocity and pressure rising rate were also decreased obviously with increasing mist concentration. Xu et al. [9] studied the inhibition effect of ultrafine water mist on methane/coal dust hybrid explosions in the presence of obstacles and pointed out that the number, shape and installation position of obstacles were the main factor for the hybrid explosions. And the explosion pressure and temperature could be obviously reduced adding mist. Jiang et al. [10] researched the effect of water mist containing phosphorus compounds on methane/coal dust explosion and found that mist as a thermal barrier could effectively reduce the flame temperature. Meanwhile, DMMP water mist could also reduce flame velocity and cut off flame propagation. And the concentrations of toxic and combustible gases could be effectively decreased during the hybrid explosions.

To effectively decrease the losses caused by mixture explosions and reduce its harm to industrial production, the venting technology as an effective method is widely adopted in petrochemical equipment [11]. The intensities of gas/dust two-phase explosions are greater than that of the single-phase substance and affected by numerous factors, resulting in a greater threat to industrial production [12]. Hence, the research on the venting and its prevention of gas/dust explosions have become a hotspot [13, 14]. Currently, scholars had conducted the venting experiment. For gas explosion venting, Zhang [15] studied the venting process of methane explosion in rectangular pipe, and found that the methane concentration had an obvious impact on the flame propagation and pressure rise during venting. Pan [16] researched the effect of venting port shape on the explosion venting, and pointed out that the square and circular venting ports with equal area had the same impact on explosion venting. However, Wang and Zhang [17, 18] proposed the diameter of venting port had an evident impact on the pressure and a smaller venting port could exacerbate the intensity of explosion venting. In addition, Ma [19] found that the ignition position had an evident impact on the propagations of pressure wave and venting flame. As the distance between the ignition source and venting port increased, the pressure emerged a variation regular of firstly reducing and then increasing. Zhang [20] also found that the venting pressure and ignition position existed an obvious correspondence, and an evident secondary venting process emerged outside the vessel as the gas was ignited at the center of vessel. Meanwhile, Li [21] also proposed that the ignition position could affect the explosion intensity of the gas, further affecting the venting pressure. The ignition at the center of vessel had the most significant impact on the explosion intensity during the venting, and it was found that the pressure history of vessel interior exhibited the multiple peaks [22]. Liu and Song [23, 24] also found that the pressure history exhibited varying degrees of oscillation and it was related to Helmholtz and acoustic oscillations. Rui [25] also observed the oscillation phenomenon of pressure history inside the venting vessel during venting, and its duration was decreased as the rupture pressure enhanced. Cao [26] found that the pressure peak value inside the vessel presented an upward changing trend with the increase of film strength. Sun [27] proposed that the pressure history appeared an evident negative value during the oscillation stage and it was related to the inertia of venting gas flow. Besides, relevant scholars found that the flame experienced a change process of "mushroom" and "spherical" structures during venting, which was mainly affected by thermodynamics and fluid mechanics [28]. Additionally, a secondary venting process appeared outside the vessel, which was related to the concentration of residual mixture inside the vessel during extinguishing stage [29].

For dust explosion venting, Marc [30] studied the effect of turbulence intensity on the venting process of dust explosion, and pointed out that its venting pressure and rising rate could be enhanced with the increase of turbulence intensity. Scheid [31] also proposed that the pressure continued to improve as the enhancement of dust turbulence. Yang [32] found that the surface area and propagation distance of venting flame showed a changing regular of firstly increasing and then reducing as the aluminum dust concentration improved. Chen and Guo [2, 33] proposed that dust concentration was also an important parameter for explosion venting. And venting pressure and flame velocity were dramatically enhanced as the dust concentration improved. Nevertheless, the venting parameter presented a downward trend after reaching the certain value. Luo [34] studied the impact of magnesium–aluminum alloy dust diameter on the venting characteristics and proposed that the metal dust with smaller diameter would produce the higher venting pressure. Gao and Huang [35, 36] also found that the change in dust particle size had a greatly impact on the explosion venting. As the dust diameter decreased, the explosion reaction rate was faster and its intensity was also more violent. Denkevits [37] proposed that the influence of dust particle size on the rate of pressure rise was more significant compared to the pressure, and it was evidently decreased with the increase of dust diameter. Besides, Yang and Chen et al. [38, 39] found that the pressure history exhibited a multi peak phenomenon by changing parameters such as ignition position and rupture pressure. Richard [40] proposed that the position where the pressure peak value outside the vessel appeared continuously moved away from the venting port as the vessel volume increased. Holbrown [41] pointed out that a larger venting port would decrease the pressure outside the vessel, and the unburned dust carried by the shock wave could be ignited and formed a bright “spherical” flame. Yan [42] found that the larger venting port and rupture pressure could enhance the probability of secondary venting. Besides, Taveau [43] also proposed that the likelihood of secondary venting appearance was greater under the higher the film pressure and dust concentration.

Totally, the inhibiting and venting as two explosion protection technology were widely adopted and could effectively protect the petrochemical equipment. The research on inhibition of gas and dust hybrid explosions by ultrafine water mist was less. The current research mainly focused on the inhibition of single flammable substance explosion by ultrafine water mist. Grasping the changes in the flame propagation characteristics and explosion parameters under the different mist conditions and obtaining amounts of accurate experimental data are very important for understanding the inhibition mechanism of gas/dust hybrid explosions. Besides, the exploration on the venting dynamics process of gas/dust two-phase explosions was deficient. And the venting process included the coupling effect such as the multiphase reaction and interphase heat-mass transfer, as well as the turbulence, and affected by numerous influencing factors. And it was more dangerous and complex than the single-phase explosion. And its venting theory was imperfect. Hence, the research on the influences of coal dust parameters on the venting parameters of gas/dust explosions is crucial for practical application and mechanism revelation.

References

1. Wang HY, Zhang YW, Xu JG, Li JL, Fan C, Hu Y. Experimental study on effect of dilute coal dust on gas explosion pressure/flame evolution process. Powder Technol 2022; 404: 117450.
2. Guo CW, Shao H, Jiang SG, Wang YJ, Wang K, Wu ZY. Effect of low-concentration coal dust on gas explosion propagation law. Powder Technol 2020; 367: 243–252.
3. Li HT, Zhai F, Li SS, Lou RY, Wang FC, Chen XK, Shu CM, Yu MG. Macromorphological features and formation mechanism of particulate residues from methane/air/coal dust gas-solid two-phase hybrid explosions: An approach for material evidence analysis in accident investigation. Fuel 2022; 315: 123209.
4. Li HT, Li SS, Zhai F, Wang FC, Lou RY, Chen XK, Shu CM, Yu MG. Effect of the initial oxidized status of coal dust on the deflagration severities and flame behaviors of pulverized coal explosion in various methane-air atmospheres. Fuel 2022; 315: 123211.
5. Ajrash MJ, Zanganeh J, Moghtaderi B. The flame deflagration of hybrid methane coal dusts in a large-scale detonation tube (LSDT). Fuel 2017; 194: 491–502.
6. Gan B, Li B, Jiang HP, Bi MS, Gao W. Suppression of polymethyl methacrylate dust explosion by ultrafine water mist/additives. J Hazard Mater 2018; 351: 346–355.
7. Feng X, Ren JJ, Pu MZ, Chen B, Bi MS. Suppression effect of ultrafine water mist on methane-coal dust hybrid explosions. Powder Technol 2022; 406: 117590.
8. Chen B, Feng X, Zhang HT, Wu ZY, Chai YX, Ren JJ, Bi MS. Experiment study on suppression of methane-coal dust compound explosion by ultrafine water mist. Fire Sci Technol 2021; 40: 1046–1051.
9. Xu HL, Wang XX, Li Y, Zhu P, Cong HY, Qin WX. Experimental investigation of methane/coal dust explosion under influence of obstacles and ultrafine water mist. J Loss Prev Process Ind 2017; 49: 929–937.
10. Jiang HP, Bi MS, Huang L, Zhou YH, Gao W. Suppression mechanism of ultrafine water mist containing phosphorus compounds in methane/coal dust explosions. Energy 2022; 239: 121987.
11. Wang CH, Guo J, Wang HZ, Zhang HW, Wu JH. Duct-vented explosion of stoichiometric hydrogen/methane/air mixtures at various hydrogen ratios. Fuel 2023; 333: 126376.
12. Kundu SK, Zanganeh J, Eschebach D, Moghtaderi B. Explosion severity of methane-coal dust hybrid mixtures in a ducted spherical vessel. Powder Technol 2018; 323: 95–102.
13. Zhao P, Tan X, Schmidt M, Wei A, Huang W, Qian X, Wu D. Minimum explosion concentration of coal dusts in air with small amount of CH_4/H_2/CO under 10-kJ ignition energy conditions. Fuel 2020; 260: 116401.
14. Zhang Q, Chen GH, Huang XJ, Xu QM, Ma JJ, Xie ML. Effects of vent layout and vent area on dynamic characteristics of premixed methane-air explosion in a tube with two side-vented ducts. J Loss Prev Process Ind 2023; 83: 105051.
15. Zhang Q, Chen GH, Xue YZ, Xu QM, Xie ML. Effects of methane volume fractions and vent areas on dynamic characteristics of vented methane-air explosion in a half-open duct. Fuel 2022; 319: 123762.
16. Pan CY, Wang XS, Li GC, Liu YP, Jiang Y. Influences of a square and circular vent on gasoline vapor explosions and its propagation: Comparative experimental study. Case Stud Therm Eng 2021; 27: 101225.
17. Wang QH, Jin SL, Luo ZM, Dai AP, Wang QF, Li ZH. Flame propagation characteristics of methane explosion under different venting conditions. Fuel 2023; 334: 126721.
18. Zhang SH, Zhang Q. Effect of vent size on vented hydrogen-air explosion. Int J Hydrogen Energy 2018; 43: 17788–17799.

19. Ma C, Wang ZR, Cui YY, Ma WD. Effect of ignition position on methane explosion in spherical vessel with a pipe. Proced Eng 2018; 211: 538–545.
20. Zhang CC, Li Y, Sun XX, Chen XF, Huang CY, Yuan BH. Investigation on the characteristics of vented stoichiometric methane-air explosion: Effect of ignition positions and obstacles. Fuel 2022; 329: 125417.
21. Li JL, Wang XB, Guo J, Zhang JQ, Zhang S. Effect of concentration and ignition position on vented methane-air explosions. J Loss Prev Process Ind 2020; 68: 104334.
22. Luo XJ, Wang CJ, Rui SC, Wan Y, Zhang Z, Li Q. Effects of ignition location, obstacles, and vent location on the vented hydrogen-air deflagrations with low vent burst pressure in a 20-foot container. Fuel 2020; 280: 118677.
23. Liu W, Guo J, Zhang JQ, Zhang S. Effect of vent area on vented deflagration of hydrogen-methane-air mixtures. Int J Hydrogen Energy 2021; 46: 6992–6999.
24. Song XZ, Zhang J, Liu Q, Xu QM, Chen J, Zhang D, Xie LF, Li B. Visualization of the external flow field during a vented explosion for hydrogen-air mixtures: Effects of hydrogen concentrations and vent areas. Fuel 2022; 327: 124965.
25. Rui S, Guo J, Li G, Wang C. The effect of vent burst pressure on a vented hydrogen-air deflagration in a 1 m^3 vessel. Energy 2018; 43: 21169–21176.
26. Cao Y, Li B, Gao KH. Pressure characteristics during vented explosion of ethylene-air mixtures in a square vessel. Energy 2018; 151: 26–32.
27. Sun S, Qiu YY, Xing HD, Wang MY. Effects of concentration and initial turbulence on the vented explosion characteristics of methane-air mixtures. Fuel 2020; 267: 117103.
28. Wang SM, Zhao Y, Li GQ, Xie YL, Wu DJ. Effect of different vent covers on the overpressure and flame propagation characteristics of hydrocarbon fuel-air mixture venting explosion. Fuel 2022; 324: 124620.
29. Zhang Y, Chen RK, Zhao MK, Luo JW, Feng WS, Fan WQ, Tan YX, Cao WG, Shu CM, Yu CJ. Hazard evaluation of explosion venting behaviours for premixed hydrogen-air fuels with different bursting pressures. Fuel 2020; 268: 117313.
30. Marc S, Andreas G, Ulrich K. Experiments on the influence of pre-ignition turbulence on vented gas and dust explosions. J Loss Prev Process Ind 2005; 2: 19.
31. Scheid M, Geißler A, Krause U. Experiments on the influence of pre-ignition turbulence on vented gas and dust explosions. J Loss Prev Process Ind 2006; 19: 194–199.
32. Yang MR, Jiang HP, Chen XF, Gao W. Characteristic evaluation of aluminum dust explosion venting with high static activation pressure. Process Saf Environ Prot 2021; 15: 83–96.
33. Chen XF, Gao W, Yang MR, Jiang HP, Zhang ZL, Zhang K. Experimental investigation on PMMA dust explosion venting at elevated static activation pressure. Powder Technol 2021; 392: 402–411.
34. Luo JW, Wang QH, Chung YH, Vikranth V, Shu CM, Cheng YC. Hazard evaluation, explosion risk, and thermal behaviour of magnesium-aluminium alloys during the polishing process by using a 20-L apparatus, MIEA, and TGA. Process Saf Environ Prot 2021; 153: 268–277.
35. Gao W, Yu JL, Li J, Zhang Q, Xie QF, Zhang XY, Hu DP. Experimental investigation on micro- and nano-PMMA dust explosion venting at elevated static activation overpressures. Powder Technol 2016; 301: 713–722.
36. Huang L, Jiang HP, Gao W. Effect of particle size and dust layer size on ignition characteristics of PMMA dust layer on hot surface. J Loss Prev Process Ind 2021; 71: 104515.
37. Denkevits A, Dorofeev S. Explosibility of fine graphite and tungsten dusts and their mixtures. J Loss Prev Process Ind 2006; 19: 174–180.
38. Yang K, Chen YB, Xiao QP, Pang L. Influence of venting coefficient on disastrous effects of aluminium powder explosions. Process Saf Environ Prot 2021; 156: 72–88.
39. Chen XF, Xu JT, Jiang HP, Gao W. Effects of ignition location on the characteristics of dust explosion venting at elevated static activation pressure. Powder Technol 2023; 420: 118408.
40. Richard S. Explosion venting technology. J Loss Prev Process Ind 1996; 9: 81–90.
41. Holbrow P, Hawksworth SJ, Tyldesley A. Thermal radiation from vented dust explosions. J Loss Prev Process Ind 2000; 13: 467–476.

42. Yan XQ, Yu JL, Gao W. Flame behaviors and pressure characteristics of vented dust explosions at elevated static activation overpressures. J Loss Prev Process Ind 2015; 33: 101–108.
43. Taveau J. Correlations for blast effects from vented dust explosions. J Loss Prev Process Ind 2010; 23(1): 15–29.

Chapter 2
Research on the Characteristics and Influence Factor of Methane and Coal Dust Two-Phase Mixture Explosions

2.1 Experimental Apparatus and Procedures

Figure 2.1 shows a schematic diagram of experiment apparatus for studying the characteristics and influence mechanism of gas/dust mixture explosions [1, 2]. It mainly included eight subsystems. The main structure had been introduced in the previous work [3], and only part of the structure was slightly modified. The powder injection system consisted of a Hartmann powder spraying device, a globe valve and a solenoid valve. The Hartmann powder spraying device was installed at the center of the bottom flange through a threaded connection. The premix tank was sequentially connected with a globe valve and a solenoid valve by the bottom end tube bundle. The coal dust was placed in the groove of the Hartmann powder spraying device. A bottom-up vortex was formed as the high-speed premixed gas flow passed through the Hartmann powder spraying device from the bottom tube bundle due to the obstruction of the hemispherical "mushroom cap", lifting the dust in the groove and forming a dust cloud. The powder spraying method could achieve the relatively uniform dust cloud. The mixture gas of methane (Purity: 99.99%) and air was prepared inside the premixed tank by the partial pressure method and the determined vacuum degree inside the vessel was achieved by a vacuum pump. The powder injection pressure and time were determined by the calculation and ensured the methane concentration inside the vessel after power spraying. And the initial pressure was normal.

The ignition system was composed of a high voltage converter, an AC contactor, a solid-state relay and an ignition electrode. The ignition electrode was placed at a central position with 110 mm height distance from the bottom flange (ignition voltage (U) = 8 kV). A high-speed camera (Photron V1212 (Japan)) was adopted to collect the flame propagation process of gas/dust mixture explosions. And flame propagation velocity history can be computed by the derivation of flame front position over time. To ensure the accuracy of flame propagation velocity, the acquisition frequency was set to 2000 fps. Namely, the change in flame front for 0.5 ms. A high-frequency

Z. Wang and X. Cao, *Mechanism of Composite Explosion Inhibition and Venting Protections*, https://doi.org/10.1007/978-981-95-4809-5_2

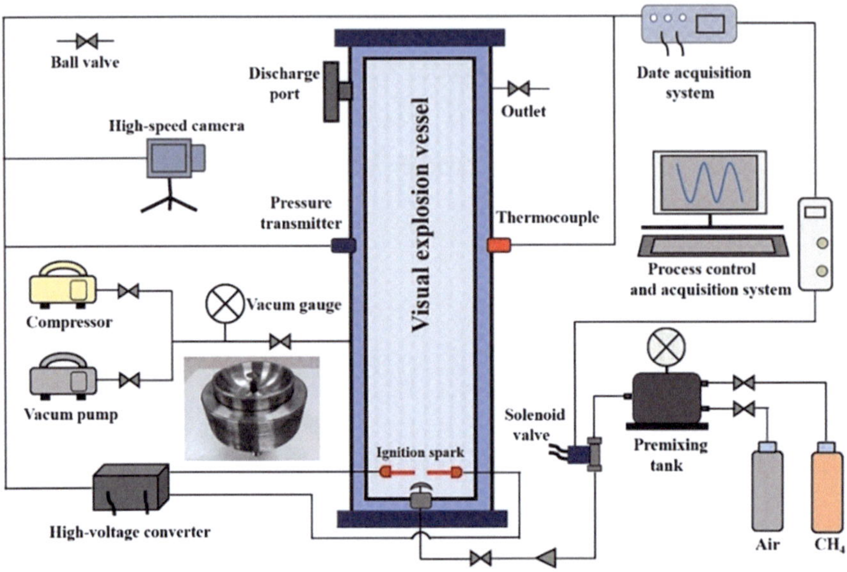

Fig. 2.1 Schematic diagram of gas/dust mixture explosion experiment apparatus

pressure transmitter (Accuracy: 0.5% FS; Acquisition frequency: 200 Hz) and a high-speed thermocouple (S-type platinum rhodium; Range: 0–1800 °C) were installed at the center of vessel sidewall to record the pressure and flame temperature of mixture explosions. The order of powder injection and ignition, as well as data collection, were realized through a program control and data collection system with a single channel of 2 M/s.

Before the experiment, the large coal block was pulverized by a grinder. And the coal dust with different particle size range was obtained by changing the screen mesh of different pore diameter. Three kinds of coal dust diameters ($d_{50} = 35.6$, 76.6 and 141.4 μm), five kinds of coal dust concentrations (q = 100, 200, 300, 400 and 500 g/m^3) and five kinds of methane concentrations (c = 2, 4, 6, 8 and 10%) were selected for studying mixture explosion experiments. Figure 2.2 shows the columnar distribution and microstructure of three coal dust diameters. The morphology of the three types of coal dusts presented a uniform state and their diameter histories were approximately normal distribution [4]. Meanwhile, the component content of coal dust was measured by industrial analysis ($M_{ad} = 8.8\%$, $A_{ad} = 6.3\%$, $V_{ad} = 42.8\%$, $FC_{ad} = 42.1\%$). Simultaneously, the pyrolysis characteristics of coal dust under the high temperature with a heating rate of 10 K/min was tested. It underwent three stages: the heat absorption, weight loss and stability. According to the comparison of explosion experiments under different powder injection pressures (P_{st}), the pressure was maximum as the P_{st} was 0.40 MPa. Hence, 0.40 MPa was selected as the powder injection pressure during experiments. Each condition was repeated 3–4 times to guarantee the experimental accuracy. The data deviation was an important parameter

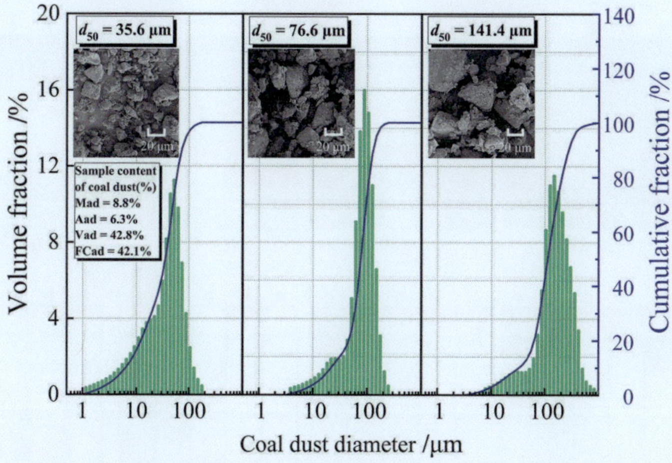

Fig. 2.2 Columnar distribution of coal dust diameter

characterizing experimental repeatability and it was calculated by the ΔP_{max} and the average value of repetitive experiments. During this experiment, the deviation value of ΔP_{max} was less than 5.5% through the data analysis.

2.2 Results and Discussions

2.2.1 Flame Propagation Characteristics

Figure 2.3 shows the comparison of mixture explosion flame propagation processes under five coal dust concentrations ($d_{50} = 76.6$ μm; $c = 8\%$). The height of flame front at the same moment showed a change trend of increase firstly and then decrease as the coal dust concentration increased. As $q = 300$ g/m^3, the flame front was the highest and brightest. The initial flame exhibited an irregular structure. Meanwhile, the flame got dark and the irregular flame became more significant with the increase and decrease of coal dust concentration. As $100 < q < 300$ g/m^3, the quantity of coal dust per unit space was gradually increased. The heat released by methane explosion could be effectively absorbed by coal dust. Subsequently, the explosion intensity was increased due to the heat release from coal dust combustion. As the flame propagated, the flame front became the smooth and developed upward in an "elliptical" structure [5]. As $q = 300$ g/m^3, the heat release of gas/dust two phase reactions after heat absorption resulted in that the explosion reaction was the most severe and the flame was the brightest. As $400 < q < 500$ g/m^3, the brightness was successively darkened and the flame began to become irregular. Especially the change was more significant as $q = 500$ g/m^3. It was because that the reaction heat of methane explosion could

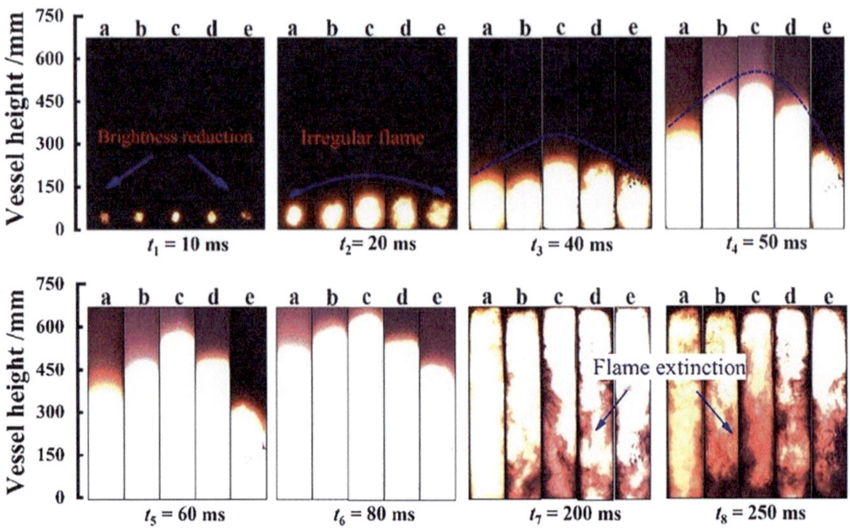

Fig. 2.3 Comparison of flame propagation processes under five coal dust concentrations (a = 100 g/m^3, b = 200 g/m^3, c = 300 g/m^3, d = 400 g/m^3 and e = 500 g/m^3)

be evidently aborted by the coal dust of higher concentration per unit space. And the coal dust could not sufficiently occur the explosion reaction due to the limited oxygen concentration inside the confined space. Meanwhile, the combustion products could greatly consume the explosion reaction heat, leading to an evident reduction in flame brightness. And the flame front exhibited an evident wrinkle and instability. This explains that the dust concentration had a significant impact on the flame structure and its propagation characteristics of gas/dust mixture explosions [6].

Figure 2.4 shows the comparison of velocity histories under different coal dust concentrations (d_{50} = 76.6 μm; c = 8%). The velocity history and its rising rate appeared a change trend of increase firstly and then decrease with the increase of coal dust concentration. As q = 300 g/m^3, the corresponding velocity value was largest and the required time for reaching the top end was the shortest. The v_{max} was evidently reduced from 17.7 to 12.9 m/s and 14.6 m/s with the change of coal dust concentration. Especially v_{max} was the smallest as q = 500 g/m^3. Figure 2.5 presents the effect of coal dust diameter on the velocity history of mixture explosions (c = 8%; q = 300 g/m^3). The velocity history was gradually reduced with the increase of coal dust diameter. In particular, v_{max} was reduced from 20.1 to 17.1 m/s and its corresponding moment was also obviously delayed. It was because that the larger coal dust possessed a smaller specific surface area, further resulting in a relatively smaller contact area with oxygen and exhibited a slower explosion reaction rate and a darker flame structure.

Figure 2.6 illustrates the influence of methane concentration on the velocity history of mixture explosions (d_{50} = 76.6 μm; q = 300 g/m^3). As the methane concentration increased, the velocity history was also appeared an alteration trend of increase firstly

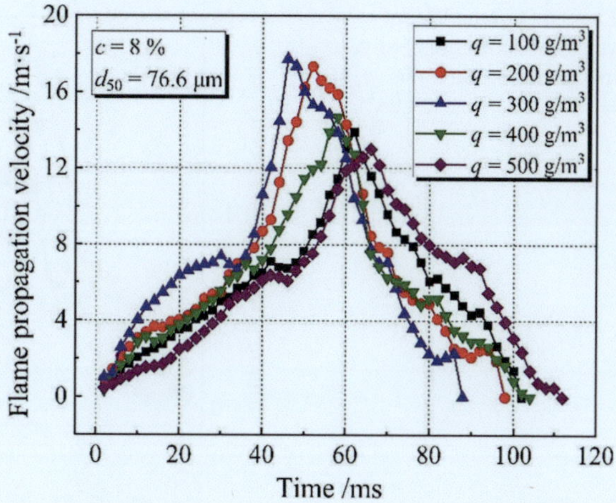

Fig. 2.4 Comparison of velocity histories under five coal dust concentrations (d_{50} = 76.6 μm; c = 8%)

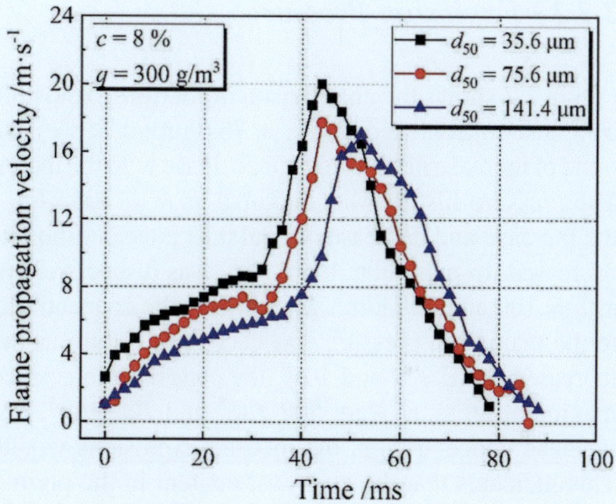

Fig. 2.5 Comparison of velocity histories under three coal dust diameters (c = 8%; q = 300 g/m^3)

and then decrease. Under 6% methane concentration, the velocity value was the maximal (v_{max} = 21.5 m/s) and the corresponding moment of v_{max} was the smallest. With the increase and decrease of methane concentration, the velocity history was evidently reduced. In particular, the velocity history was smallest (v_{max} = 12.4 m/s) and the corresponding moment of v_{max} was increased from 44 to 82 ms as c = 10%. This was because that the methane and coal dust could not obtain sufficient oxygen during the explosion reaction as the methane concentration was increased to 10%, resulting in the inability of the heat released by the mixture system to maintain a high flame propagation velocity. Based on the Fig. 2.6, the methane concentration and coal dust parameters could evidently affect the flame propagation characteristics of gas/dust two-phase mixture explosions.

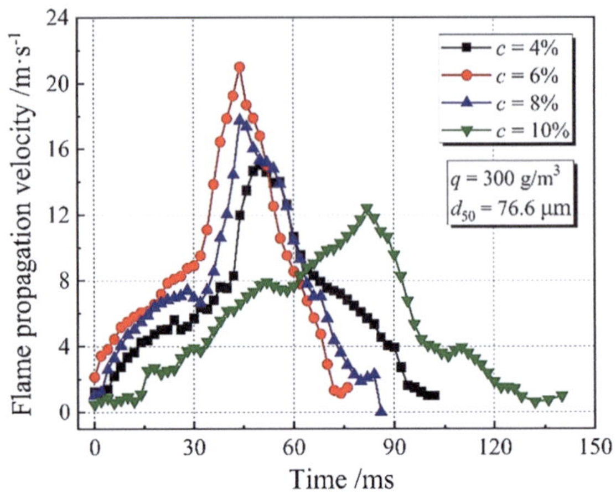

Fig. 2.6 Comparison of velocity histories under five methane concentrations (d_{50} = 76.6 μm; q = 300 g/m^3)

2.2.2 *Explosion Pressure*

Figure 2.7 shows the comparison of mixture explosion pressure histories under five coal dust concentrations (d_{50} = 76.6 μm; c = 8%). The ΔP_{max} appeared a change trend of increase firstly and then decrease with the increase of coal dust concentration. The pressure history was maximal as q = 300 g/m^3 (ΔP_{max} = 0.61 MPa). With the increase and decrease of coal dust concentration, the pressure was significantly decreased. In particular, the ΔP_{max} was respectively reduced to 0.57 and 0.52 MPa as q = 100 and 500 g/m^3. Meanwhile, the concentration of coal dust corresponding to the maximum pressure was decreased in turn as the methane concentration was increased. As c = 2 and 10%, the concentrations of coal dust corresponding to the maximum pressure were 500 g/m^3 and 200 g/m^3, respectively. Especially for 2% methane concentration, the mixture explosions would not occur as q < 400 g/m^3. This indicates that the coal dust content in the premixed gas of methane/air had a significant influence on the intensity of gas/dust two-phase mixture explosions.

Figure 2.8 presents the comparison of mixture explosion pressure histories under three coal dust diameters (q = 300 g/m^3; c = 8%). The pressure was evidently reduced as the coal dust diameter increased. In particular, ΔP_{max} was decreased by 3.17% and 11.11%, respectively. As d_{50} = 141.4 μm, the pressure value was the smallest (ΔP_{max} = 0.56 MPa). Meanwhile, ΔP_{max} all appeared a decrease change trend as the coal dust diameter increased. Besides, the intensity of mixture explosions was maximal as q = 300 g/m^3. It also indicates that not only the concentration of coal dust could affect the explosion intensity, but also its specific surface area also could evidently affect the gas/dust two-phase mixture explosions.

Figure 2.9 presents the comparison of mixture explosion pressure histories under different methane concentrations (q = 300 g/m^3; d_{50} = 76.6 μm). The pressure history also showed a trend of increase firstly and then decrease as the methane concentration increased. As c = 6%, ΔP_{max} was the maximal (ΔP_{max} = 0.63 MPa). With the increase and decrease of methane concentration, ΔP_{max} presented an obvious

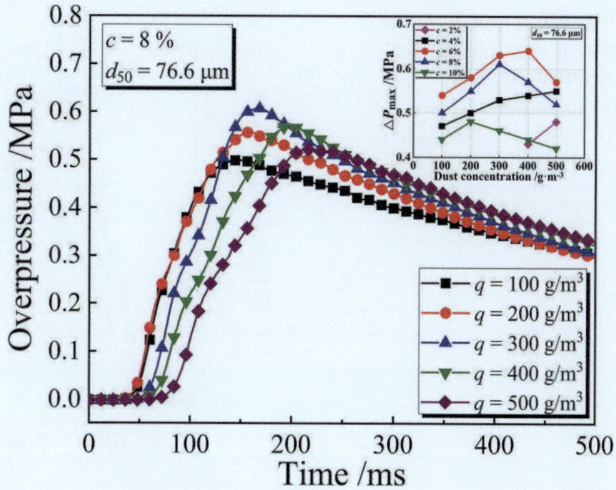

Fig. 2.7 Comparison of pressure histories under different coal dust concentrations (d_{50} = 76.6 μm; $c = 8\%$)

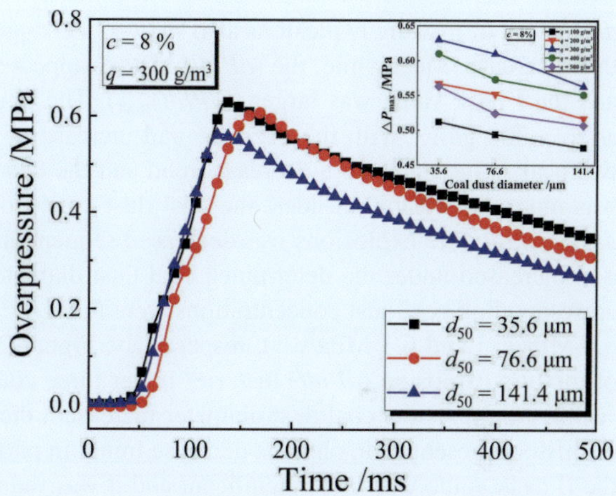

Fig. 2.8 Comparison of pressure histories under different coal dust diameters ($c = 8\%$; $q = 300$ g/m^3)

decrease trend. And the pressure all showed the same change trend under five coal dust concentrations. Especially for 2% methane concentration, the mixture explosions would not occur as $q < 400$ g/m^3. And the same phenomenon also occurred as $c > 10\%$. It explains that the concentrations of combustible gas and dust could evidently affect the explosive limit of gas/dust two phases.

2.2.3 *Explosion Pressure Rising Rate*

Figure 2.10 illustrates the comparison of (dP/dt) histories under five coal dust concentrations ($c = 8\%$; $d_{50} = 76.6$ μm). It clear that the (dP/dt) history exhibited the same change trend as the pressure history. As the coal dust concentration increased, $(dP/$

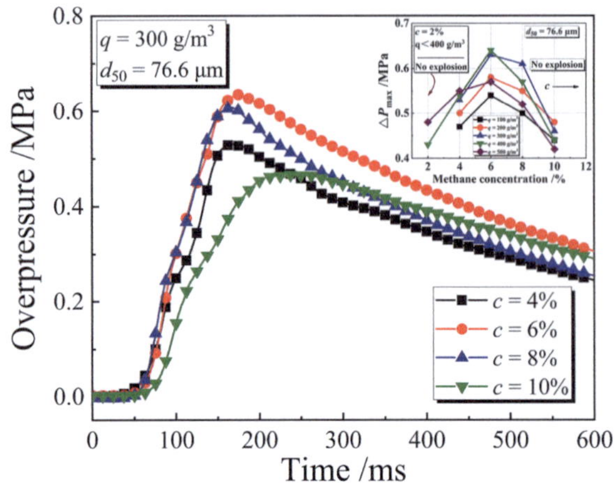

Fig. 2.9 Comparison of pressure histories under different methane concentrations ($d_{50} = 76.6$ μm; $q = 300$ g/m^3)

dt) history of mixture explosions also showed a change trend of increase firstly and then decrease. Meanwhile, the (dP/dt) history appeared two peaks (I and II peaks) and the I peak value was larger ($(dP/dt)_{max}$). The two peak values were maximal as $q = 300$ g/m^3. With the decrease and increase of coal dust concentration, the two peak values showed a decrease trend and the decrease degree of II peak value was more significant. Besides, the coal dust concentration corresponding to $(dP/dt)_{max}$ of mixture explosions was decreased sequentially as the methane concentration increased under the determined coal dust diameter. As $c = 4\%$ and 10%, the corresponding coal dust concentrations were 500 g/m^3 and 200 g/m^3 ($(dP/dt)_{max} =$ 8.4 MPa/s^{-1} and 6.1 MPa/s^{-1}), respectively. Figure 2.11 presents the comparison of mixture explosion (dP/dt) histories under three coal dust diameters ($c = 8\%$; $q = 300$ g/m^3). As the coal dust diameter increased, the I and II peak values of (dP/dt) history presented an obvious decrease trend. In particular, the decrease extent of the II peak value was more significant and it was reduced by 30.39%. Meanwhile, $(dP/dt)_{max}$ all appeared a decrease trend and the maximum reduction extent reached 46.74% with the increase of coal dust diameter as $c = 8\%$.

Figure 2.12 shows the comparison of mixture explosion (dP/dt) histories under different methane concentrations ($q = 300$ g/m^3; $d_{50} = 76.6$ μm). It clear that the two peak values of (dP/dt) history showed a trend of increase firstly and then decrease as the methane concentration was raised. As $c = 6\%$, the pressure rising rate was maximal. In addition, the $(dP/dt)_{max}$ were decreased by 25.49 and 33.33% under 4 and 6% methane concentrations. As $d_{50} = 76.6$ μm, the pressure rising rates corresponding to 6% and 10% methane concentrations were the maximal and minimal values, respectively. The above explains that the reaction rate of gas/dust two-phase mixture explosions could be evidently affected by the methane content, coal dust diameter and its concentration, thereby affecting the explosion intensity.

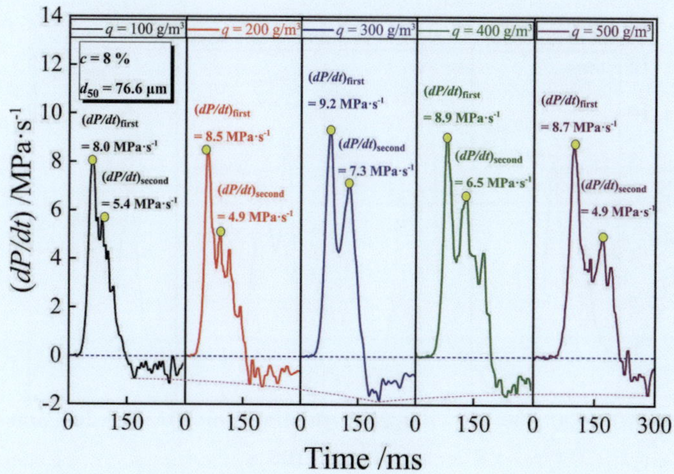

Fig. 2.10 Comparison of (dP/dt) histories under different coal dust concentrations ($d_{50} = 76.6$ μm; $c = 8\%$)

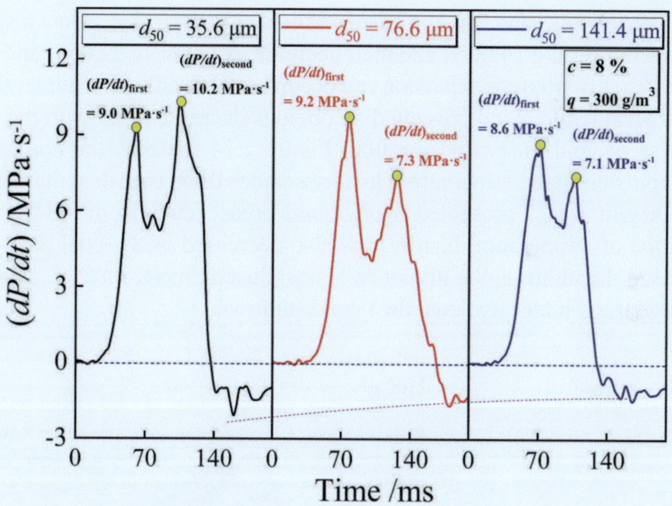

Fig. 2.11 Comparison of (dP/dt) histories under different coal dust diameters ($c = 8\%$; $q = 300$ g/m^3)

2.2.4 Flame Temperature

Figure 2.13 is the comparison of flame temperature histories under five coal dust concentrations ($c = 8\%$; $d_{50} = 76.6$ μm). The temperature also showed the same change trend as the pressure history with the increase of coal dust concentration. Especially the flame temperature was the highest and T_{max} reached 1687 °C as $q =$

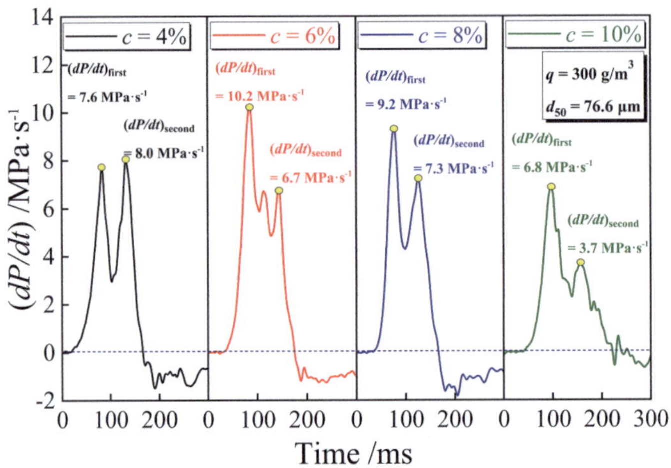

Fig. 2.12 Comparison of (dP/dt) histories under different methane concentrations ($d_{50} = 76.6$ μm; $q = 300$ g/m^3)

300 g/m^3. For 100 g/m^3 and 500 g/m^3 methane concentrations, $T_{\max}$ were respectively reduced to 1315 °C and 1438 °C and their decrease extent were 22.05% and 14.76%. As $c = 8\%$, $T_{\max}$ of mixture explosions all occurred at $q = 300$ g/m^3 under three dust diameters. Meanwhile, $T_{\max}$ presented an obvious decrease trend with the decrease and increase of coal dust concentration. Figure 2.14 presents the comparison of mixture explosion flame temperature histories under three coal dust diameters ($c = 8\%$; $q = 300$ g/m^3). $T_{\max}$ presented an obvious decrease trend (from 1781 to 1569 °C) and the slope of temperature history was also decreased as the coal dust diameter was increased. In addition, the influence of coal dust diameter on $T_{\max}$ all showed a same change trend under five coal dust concentrations.

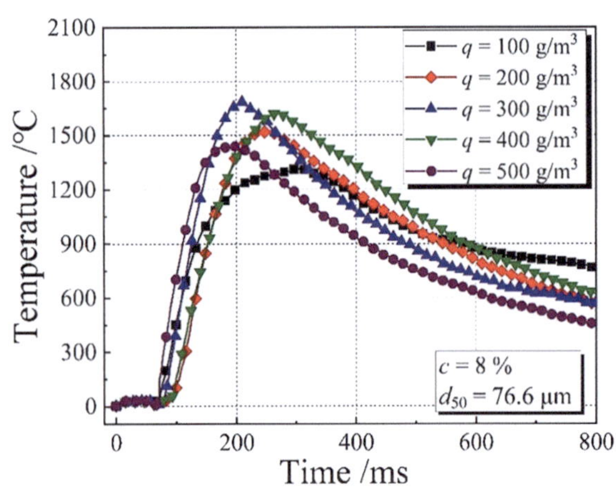

Fig. 2.13 Comparison of temperature histories under different coal dust concentrations ($d_{50} = 76.6$ μm; $c = 8\%$)

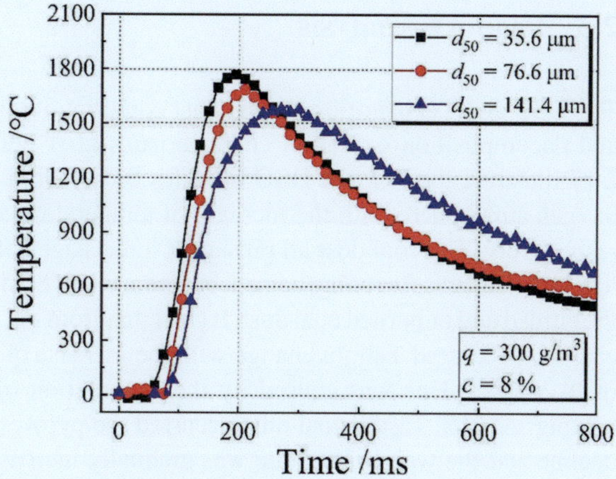

Fig. 2.14 Comparison of temperature histories under different coal dust diameters ($c = 8\%$; $q = 300$ g/m^3)

Figure 2.15 presents the comparison of mixture explosion flame temperature histories under different methane concentrations ($q = 300$ g/m^3; $d_{50} = 76.6$ μm). $T_{\max}$ of 6% methane concentration was the highest. With the change of methane concentration, the temperature history and its slope were decreased successively. Especially for 4% and 10% methane concentrations, $T_{\max}$ was decreased by 22.19% and 44.85%, respectively. Besides, the coal dust concentration corresponding to the highest flame temperature was also successively reduced with the increase of methane concentration under 76.6 μm coal dust diameter. Especially, the coal dust concentration corresponding to the highest flame temperature was 200 g/m^3 as $c = 10\%$. This indicates that the combustion degree and heat release of gas/dust two-phase could be evidently affected by the combustion gas concentration, the dust diameter and its concentration.

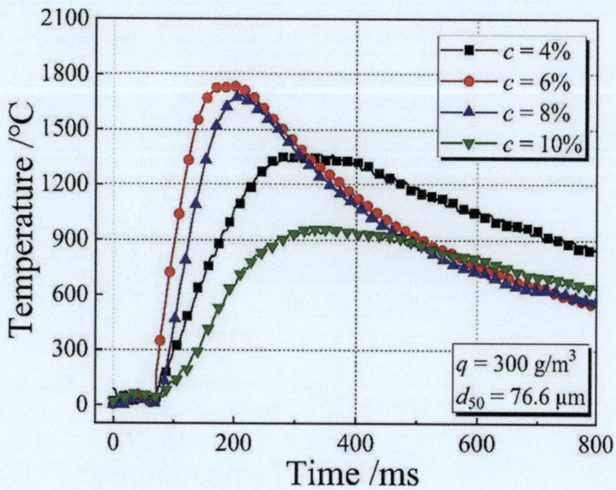

Fig. 2.15 Comparison of temperature histories under different methane concentrations ($d_{50} =$ 76.6 μm; $q = 300$ g/m^3)

2.3 Product Analysis

The DSC and TG histories of coal dust could intuitively reflect its thermal stability and decomposition oxidation characteristics after being heated. Figures 2.16 and 2.17 illustrate the TG and DSC histories of burned and unburned coal dusts in an oxygen atmosphere with the increase of temperature at a heating rate of 10 K/min, respectively. The coal dust all presented three stages of the temperature rise, weight loss and stability after being heated, but there were the significant differences between the burned and unburned coal dust. It clear that from Fig. 2.16, TG history of unburned coal dust occurred a slight change with the increase of temperature within the range of 0–263 °C. This was caused by the evaporation of moisture in coal dust after heating. At 263 °C, the coal dust occurred the pyrolysis. And TG history started to decline and the weight loss rate was gradually increased. The weight loss rate was the fastest and the corresponding DTG history reached a peak value (1.25%/°C) at 446 °C. This process was considered the acceleration weightlessness stage of coal dust. Meanwhile, the slope of coal dust DSC history was rapidly increased at this stage, indicating that the coal dust experienced a violent oxidation reaction and the heat release was continuously increased. During the whole thermal decomposition process (263 < T < 612 °C), there were two exothermic peaks on the DSC history at 446 and 484 °C, and the first peak (34.91 mW/mg) was significantly greater than the second peak (13.25 mW/mg). The appearance of the second exothermic peak resulted from the pyrolysis of stable aromatic rings inside the coal dust [7]. After the first peak value, the weight loss rate was rapidly decreased. As the temperature increased to 612 °C, the weight loss did not occur anymore and reached the thermal stability stage.

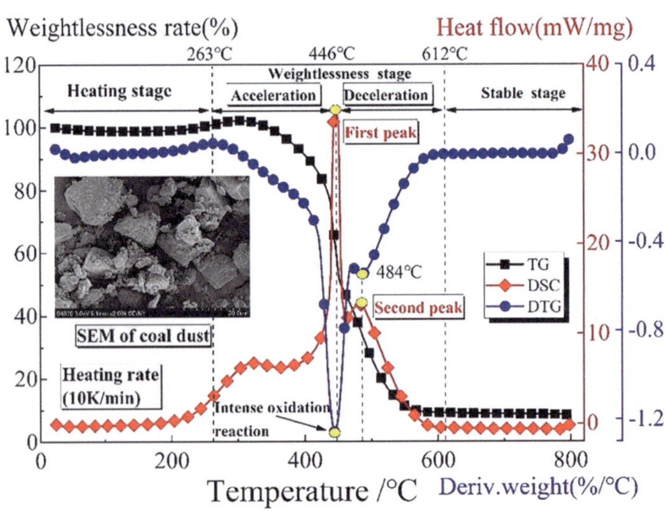

Fig. 2.16 TG and DSC histories of unburned coal dust

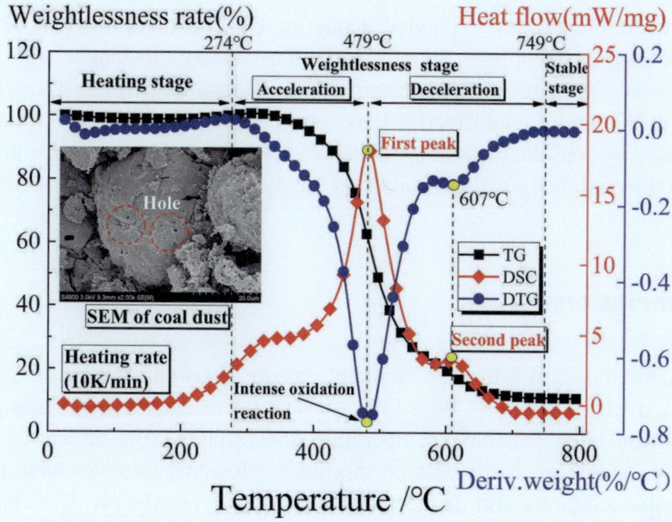

Fig. 2.17 TG and DSC histories of burned coal dust

Compared with unburned coal dust, the thermal stability and decomposition oxidation of burned coal dust occurred an obvious change, as shown in Fig. 2.17. At 274 °C, the burned coal dust appeared the pyrolysis and the corresponding temperature was slightly increased. After that, the coal dust entered the accelerated weight loss stage, but the weight loss rate was obviously decreased. Especially the peak value of weight loss rate (0.78%/°C) was evidently reduced and the corresponding temperature was increased ($T = 479$ °C). Meanwhile, the slope of DSC history during the accelerated weightlessness stage was also significantly decreased. Especially the exothermic peak value (18.18 mW/mg) of coal dust oxidation reaction was also significantly decreased and the corresponding temperature was evidently increased. Although the DSC history also appeared a small second peak, it was obviously reduced and the corresponding temperature was prolonged compared with the unburned coal dust. Besides, the temperature of coal dust reaching the stable stage was also increased. It indicates that the content of stable aromatic rings in coal dust was significantly decreased after the combustion, but there was still a small pyrolysis reaction of remaining stable aromatic rings at the high temperature.

The above indicates that coal dust participated in the gas/dust two-phase explosion reactions and the pyrolysis reaction of volatile matter resulted in an evident reduction in the weight loss and oxidation reaction rate of coal dust. It can be seen by the SEM that the unburned coal dust presented a blocky structure with sharp edges and corners. Under the action of a high temperature flame, the volatile matter in coal dust continuously occurred the pyrolysis reaction. The macromolecular chemical bonds were decomposed into gaseous state (CH_4, CO, CO_2, N_2 and O_2, etc.) and small molecular liquids (hydrocarbons) [8], resulting in a continuous decrease in the functional groups involved in the oxidation reaction. Besides, the volatile matter gradually accumulated and expanded inside the coal dust, resulting in an obvious increase of coal dust volume. Subsequently, the volatile substances were released from its surface after reaching a certain pressure, resulting in the numerous pore structure on the coal dust surface [9], as can be seen from Fig. 2.17. The volatile gas generated by coal dust would promote the gas phase explosion reaction and heat transfer to the coal dust, further promoting the pyrolysis reaction of coal dust.

2.4 Conclusions

The characteristics and influence factor of gas/dust two-phase mixture explosions were studied experimentally. Based on the qualitative analysis of flame propagation, explosion intensity, flame temperature and products, the effects of methane and coal dust physical parameters on mixture explosion parameter were obtained. And the influence factor and its mechanism were analyzed deeply. Conclusions are summarized as follows:

First, the flame propagation of mixture explosions could be evidently affected by the coal dust parameters and methane concentration. The height of flame front at the same moment showed a change trend of increase firstly and then decrease with the rise of coal dust concentration. The v_{max} also presented the same change trend. In particular, the flame front was the highest and the brightness was the strongest as $q = 300$ g/m^3. With the increase and decrease of coal dust concentration, the flame brightness was gradually darkened and the irregular flame front was more significant. The velocity history also appeared the same change trend as the methane concentration increased. And the maximal velocity value appeared at 6% methane concentration. However, the time for reaching the vessel end was increased and v_{max} was gradually decreased as the coal dust dimeter increased.

Second, ΔP_{max} all appeared a change trend of increase firstly and then decrease with the increases of methane and coal dust concentrations. The coal dust concentration corresponding to the maximum pressure existed difference and was decreased in turn with the rise of methane concentration. As $c = 2$ and 10%, the corresponding coal dust concentrations were 500 g/m^3 and 200 g/m^3, respectively. However, ΔP_{max} was obviously reduced with the rise of coal dust diameter. The (dP/dt) history also showed the similar changing trend to the pressure with the changes of methane and coal dust concentrations and appeared two peak values. With the rise of coal dust diameter, the two peak values all showed a decrease trend. And the reduction extent of second peak value was more significant.

Third, T_{max} presented the similar changing trend to the pressure with the increases of methane and coal dust concentrations (increase firstly and then decrease). However, T_{max} showed a decreasing change trend as the coal dust diameter raised. Besides, the coal dust participated in the gas/dust two-phase explosion reaction at high temperature. The thermal stability and decomposition oxidation characteristics of burned coal dust appeared an obvious change compared with the unburned coal dust. The pyrolysis reaction of volatile matter led to an obvious reduction in the weight loss and oxidation reaction rate. And the precipitation of volatile matter also resulted in an obvious pore structure on its surface.

For better prevention of gas/dust two-phase mixture explosions, three suggestions are proposed: (1) The coal dust concentration corresponding to the maximum pressure was different with the change of methane concentration. In practice, the mixture concentration should be kept away from the concentration corresponding to the maximum mixture explosion pressures. (2) Enterprises involved in powder production should avoid the presence of small size dust as much as possible and the mixture of the two should exceed its explosion limit. (3) The space where dust existed should be sprayed with fine water mist to decrease the precipitation of volatile matter inside the dust and its combustion rate.

References

1. Cao XY, Lu YW, Jiang JC, Wang ZR, Wei HY, Li YM. Experimental study on explosion inhibition by heptafluoropropane and its synergy with inert gas. J. Loss Prev. Process Ind. 2021; 71: 104440.
2. Cao XY, Bi MS, Ren JJ, Chen B. Experimental research on explosion suppression affected by ultrafine water mist containing different additives. J. Hazard. Mater. 2019; 368: 613–620.
3. Cao XY, Ren JJ, Zhou YH, Wang QJ, Gao XL, Bi MS. Suppression of methane/air explosion by ultrafine water mist containing sodium chloride additive. J. Hazard. Mater. 2015; 285: 311–318.
4. Castellanos D, Carreto-Vazquez VH, Mashuga CV, Trottier R, Mejia AF, Mannan MS. The effect of particle size polydispersity on the explosibility characteristics of aluminum dust. Powder Technol. 2014; 254: 331–337.
5. Chen XF, Hou ZX, Zhao Q, Li Q, Li Y, Huang CY, Dai HM. Suppression of methane/coal dust deflagration by $Al(OH)_3$ based on flame propagation characteristics and thermal decomposition. Fuel 2022; 311: 122530.
6. Zhang L, Wang HY, Chen C, Wang PP, Xu LW. Experimental study to assess the explosion hazard of CH_4/coal dust mixtures induced by high-temperature source surface. Process Saf. Environ. Prot. 2021; 154: 60–71.
7. Wang JH, Du J, Chang LP, Xie KC. Study on the structure and pyrolysis characteristics of Chinese western coals. Fuel Process Technol. 2010; 91: 430–433.
8. Liu ZT, Zhang SS, Xi RZ, Guo RL, Liu BB, Lin S. Analysis on residual gas characteristics of coal dust explosion in confined space. J. Coal Sci. 2015; 40: 1574–1579.
9. Nie BS, Liu XF, Yang LL, Meng JQ, Li XC. Pore structure characterization of different rank coals using gas adsorption and scanning electron microscopy. Fuel 2015; 158: 908–917.

Chapter 3
Research on the Inhibition of Methane/Coal Dust Hybrid Explosions by the Ultrafine Water Mist

3.1 Experimental Apparatus and Procedures

Figure 3.1 shows a schematic diagram of experimental apparatus for inhibiting methane and coal dust hybrid explosions by ultrafine water mist inside the closed visualization vessel. It mainly includes a visual explosion vessel, an ignition system, a pressure acquisition system, a temperature acquisition system, an image acquisition system, a gas distribution system, an ultrasonic atomization system, a powder spraying system, a program control and data acquisition system, etc.

The experimental apparatus was introduced in the previous work. In this experiment, the apparatus was slightly altered. Namely, the gas distribution system was composed of a gas supply source (the methane and air), a premix tank and a pressure gauge. The premixed gas was prepared in the premix tank by the Dalton's law of partial pressure. The explosion vessel was a rectangular vessel (910 mm × 150 mm × 150 mm). Two tempered glasses with good optical characteristic were installed in the front and back sidewalls of the vessel to record the flame propagation process. The mist was produced by the ultrasonic atomization method and the atomization system was placed in the top end of vessel by the fixed plate. Once connected the power, the mist was generated in the vessel directly. The atomization rate ($q_{\mathrm{m}} = 1.146$ g/m^3) was measured using a high-precision electronic balance with 0.001 g accuracy and the mist diameter was measured by a phase Doppler particle analyzer ($D_{32} = 10.03$ μm). With the mist amount increased, the mist entered the explosion vessel through the atomization cup outlet and uniformly dispersed in the vessel under the combined action of gravity and drag forces. The mist concentration was gradually increased with increasing atomization time. The water mist concentration was calculated by the ratio of water mist amount to vessel volume. The dust spraying system was consisted of the solenoid valve and the Hartmann device, as shown in Fig. 3.1. The coal dust of certain mass and diameter was placed in the Hartmann dust groove which connected with the premix tank through the solenoid valve. The opening and closing of solenoid valve was controlled by the program and data acquisition system, further realizing

Z. Wang and X. Cao, *Mechanism of Composite Explosion Inhibition and Venting Protections*, https://doi.org/10.1007/978-981-95-4809-5_3

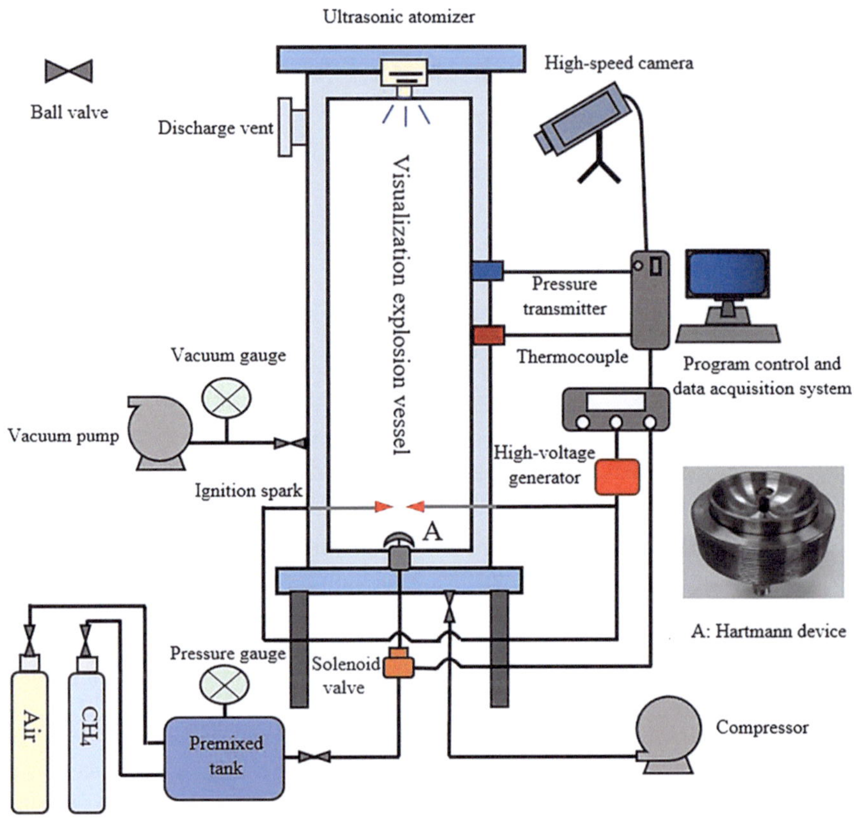

Fig. 3.1 Schematic diagram of the experimental apparatus

the control of dust spraying time. A high-speed camera (Phantom V1212) was used to record the flame propagation process of methane and coal dust hybrid explosions under different experimental conditions (the acquisition frequency: 2000 fps; the resolution: 1280 × 1024). The bright flame front could be clearly distinguished with the unburned gas and its change with an interval of 0.5 ms was obtained by the high-speed camera data processing system. And the flame front position was accurately determined by the pixel coordinate. Meanwhile, the actual size of the flame front position inside the vessel was calculated through the equal scale conversion. A high-frequency pressure transmitter (MD-HF; the response frequency: 50 kHz; the range: 0–2.5 MPa; the acquisition accuracy: 0.25%) and a S-type high-frequency thermocouple (the range: 0–3000 °C) were respectively installed in the middle of vessel wall surface to collect the explosion pressure and flame temperature with time. An eight-channel high frequency data acquisition card (USB287X) was adopted to realize the program control and data acquisition (the acquisition frequency: 200 kHz) [1].

During this experiment, three kinds of coal dust diameters with the same type were selected to carry out the inhibition of hybrid explosions. Before the experiment, the coal of large block was pulverized by a grinder. And the coal dust with different particle size range was obtained by changing the screen mesh of different pore diameter. And the diameter of coal dust was measured by a particle size tester (PT-01) ($d_{50} = 35.6$, 76.6 and 141.4 μm). The "d_{50}" represented the median diameter of coal dust. Its mean the coal dust with particle size larger than and smaller than the medium diameter accounts for 50% respectively. And the diameter presented a well normal distribution, as shown in Fig. 3.2. The composition of coal dust was analyzed by an automatic industrial analyzer (5E-MAG6700II). The internal components (i.e., the moisture, ash, volatile and fixed carbon) were determined respectively, as shown in Table 3.1. The surface structure of coal dust was measured by a scanning electron microscope (S-4800). Besides, the coal dust was dried for 48 h before the experiment. The high purity methane (99.99%) was selected to prepare the premixed gas of certain methane concentration ($c = 8\%$) and pressure in the premix tank.

Before the experiment, the coal dust with determined mass and diameter was placed in the Hartmann dust groove. And the vacuum pump was adopted to realize the certain vacuum degree inside the vessel. The premixed gas was prepared in the premixed tank. Then the coal dust was uniformly dispersed in the explosion vessel by the air flow carrying method through the opening and closing of the solenoid valve controlled by the program control and data acquisition system. The duration of powder spraying under different injection pressure was calculated. The injection pressure of 0.50 MPa could ensure the premixed gas concentration and the initial pressure (0.1 MPa) inside the vessel at the ignition moment. And the hybrid explosions were the most violent under the duration of 800 ms powder spraying. It indicates that the coal dust was evenly distributed inside the vessel. Meanwhile, it can be observed by the high-speed camera that the coal dust could reach the top end of vessel and presented a good distribution state. Five coal dust concentrations were

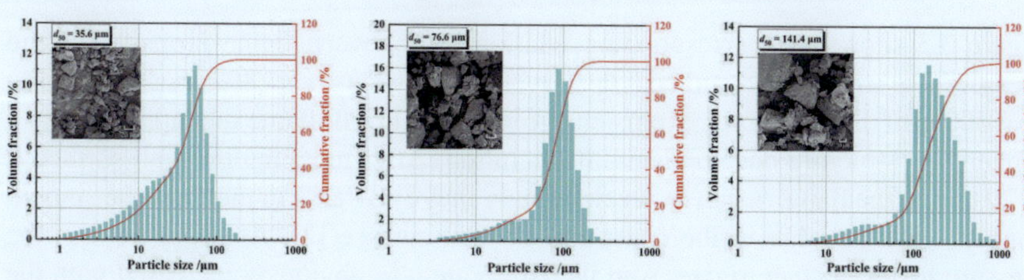

Fig. 3.2 Diameter distribution of coal dust and its surface structure ($d_{50} = 35.6$, 76.6 and 141.4 μm)

Table 3.1 Proximate analysis of coal dust (wt%)

Sample	M_{ad}	A_{ad}	V_{ad}	FC_{ad}
Coal dust (%)	8.8	6.3	42.8	42.1

M_{ad}: moisture content; V_{ad}: volatile matters; A_{ad}: ash; FC_{ad}: fixed carbon

adopted during experiments (q (coal dust concentration) = 100, 200, 300, 400, and 500 g/m^3). Namely, the ratio of coal dust mass to vessel volume. A high voltage discharge method was used to ignite the methane and coal dust (Ignition energy = 8 kV) and the ignition duration was 100 ms. The tungsten steel wire with 2 mm diameter was adopted as electrode material. The gap of two electrodes was 6 mm and the height difference with the bottom end was 110 mm. Before ignition, the determined mist concentration was realized according to the atomization rate and time. Five water mist concentrations were selected for hybrid explosions inhibition (Q (mist concentration) = 0, 141, 282, 423, and 564 g/m^3). Namely, the ratio of atomized water mist amount to vessel volume. After ignition, the premixed gas and coal dust were ignited. And the gas/dust hybrid explosions flame propagated from the bottom to top end of vessel. The coal dust combustion products were thoroughly cleaned and the gas combustion products were replaced by fresh air after the experiment. After the methane/coal dust explosion, the coal dust remaining and adhering to the vessel wall surface still had a slightly impact on the next group of experiments although the explosion vessel was wiped and cleaned. Meanwhile, a small amount of coal dust would also adhere to the wall surfaces of window and vessel during the powder spraying, resulting the slightly reduction of the dust concentration suspended inside the vessel. Hence, each condition was repeated 3–4 times to guarantee the experimental accuracy. And the deviation values of the ΔP_{max} and T_{max} were 3.4% and 4.8% respectively.

3.2 Results and Discussions

3.2.1 *Relationship Between Flame Propagation and Pressure Rise*

Figure 3.3 shows the corresponding relationship between flame propagation and pressure rise of methane and coal dust hybrid explosions ($c = 8\%$; $d_{50} = 76.6$ μm; $q = 300$ g/m^3). The hybrid explosions presented a bright flame and approximate a "spherical" structure development after ignition. After touching the wall surface ($t = 20.0$ ms), the flame began to acceleratingly move upwards in the form the smooth "finger" structure due to the constraint of wall surface [2]. As $t = 31.9$ ms, the pressure history began to rise. And its rising rate was obviously increased with the acceleration propagation of flame. As $t = 44.6$ ms, the flame propagation velocity reached the peak value ($v_{max} = 19.7$ m/s). Then the curvature radius of flame front was gradually increased as its propagated. The velocity history was obviously decreased but the flame still maintained a large propagation velocity. At this stage, the pressure appeared the first acceleration rise process ($t = 31.9$–86.0 ms) due to the acceleration propagation of flame and the pressure rising rate history appeared the first peak value ($(dP/dt)_{first}$) at 69.5 ms. Under the action of reflected pressure wave, the flame propagation velocity was obviously decreased [3]. However, the pressure rising rate history

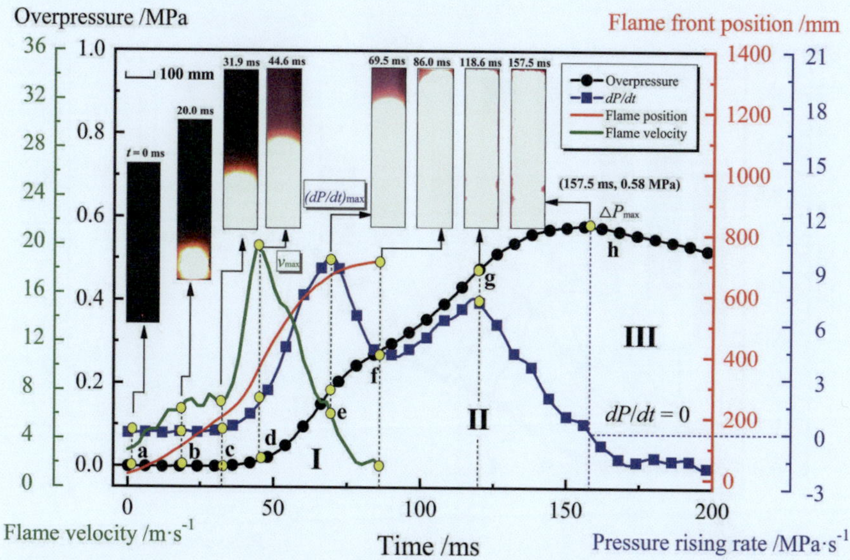

Fig. 3.3 Relationship between flame propagation and pressure rise of methane and coal dust hybrid explosions ($c = 8\%$; $d_{50} = 76.6$ μm; $q = 300$ g/m^3)

presented the second acceleration rise process ($t = 86.0$–157.5 ms) and appeared the second peak value ($(dP/dt)_{second}$) at 118.6 ms due to the sharp increase of hybrid explosions heat accumulation. At 118.6 ms, the flame reached the top end of vessel and the explosion reaction was the most violent. As $t = 157.5$ ms, the pressure history reached the peak value ($\Delta P_{max} = 0.58$ MPa). After that, the pressure history was sharply decreased as the flame was acceleratingly extinguished.

After adding mist, the flame propagation and pressure rise appeared an obvious change. Figure 3.4 shows the corresponding relationship between flame propagation and pressure rise of hybrid explosions under the action of ultrafine water mist ($c = 8\%$; $d_{50} = 76.6$ μm; $Q = 282$ g/m^3; $q = 300$ g/m^3). It can be seen that the flame presented the dark red after ignition. The flame also accelerated upward propagation, but the flame front presented an obvious irregular structure phenomenon ($t = 20.0$–42.5 ms) and did not also appeared a "spherical" structure. It was because that the flame showed an obvious instability at the initial stage due to the endothermic effect of water mist. The height of flame front was also relatively low compared with no mist condition. The time for the pressure history began to rise also was delayed ($t = 42.5$ ms). As $t = 82.7$ ms, the flame propagation velocity reached the peak value ($v_{max} = 12.3$ m/s). Compared with the no mist condition, v_{max} was obviously reduced and its appearance moment was prolonged. And the flame still presented the dark red. The flame structure became wrinkle and did not show a smooth "finger" structure. Due to the acceleration propagation of flame, the pressure appeared the first acceleration rise process ($t = 42.5$–126.0 ms). However, the acceleration rise stage was obviously delayed. The first peak value of (dP/dt) history ($(dP/dt)_{first} =$

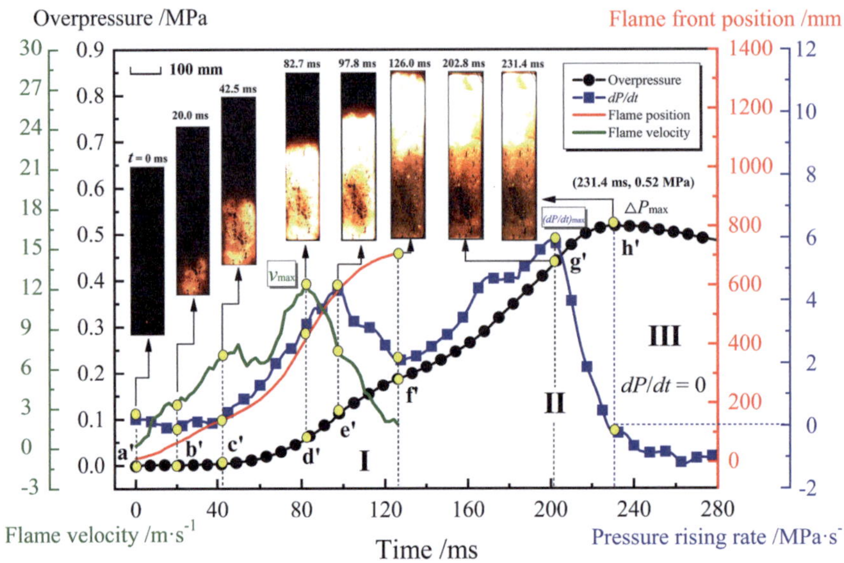

Fig. 3.4 Relationship between flame propagation and pressure rise of methane and coal dust hybrid explosions under ultrafine water mist condition ($c = 8\%$; $d_{50} = 76.6$ μm; $q = 300$ g/m^3; $Q = 282$ g/m^3)

4.3 MPa/s^{-1}) was decreased and its appearance moment was prolonged ($t = 97.8$ ms). After that, the pressure history presented the second acceleration rise process (t = 126.0–231.4 ms). And the second peak value ($(dP/dt)_{second} = 5.9$ MPa/s^{-1}) was also decreased and the its appearance moment was prolonged ($t = 202.8$ ms). Compared with the first peak, the second peak was increased obviously due to the effect of vapor pressure produced by the mist vaporization at the high temperature inside the closed vesse. Meanwhile, the pressure peak value ($\Delta P_{max} = 0.52$ MPa) was also reduced and its appearance moment was prolonged ($t = 231.4$ ms). It indicates that the addition of ultrafine water mist could effectively reduce the reaction rate and intensity of methane and coal dust hybrid explosions.

3.2.2 *Analysis of Flame Propagation Velocity*

Figure 3.5 shows the flame propagation process of methane and coal dust hybrid explosions under different mist concentrations ($c = 8\%$; $d_{50} = 76.6$ μm; $q = 300$ g/m^3). It can be seen that the flame brightness was obviously decreased and changed from the bright white to dark red with increasing mist concentration. The flame no longer presented the smooth structure and the deformation extent was obviously increased. And the change was more significant with the increase of mist concentration. It indicates that the flame front appeared an obvious thermal diffusion instability and presented an irregular structure due to the endothermic effect of ultrafine water mist [4]. Meanwhile, the height of flame front was decreased in turn and the

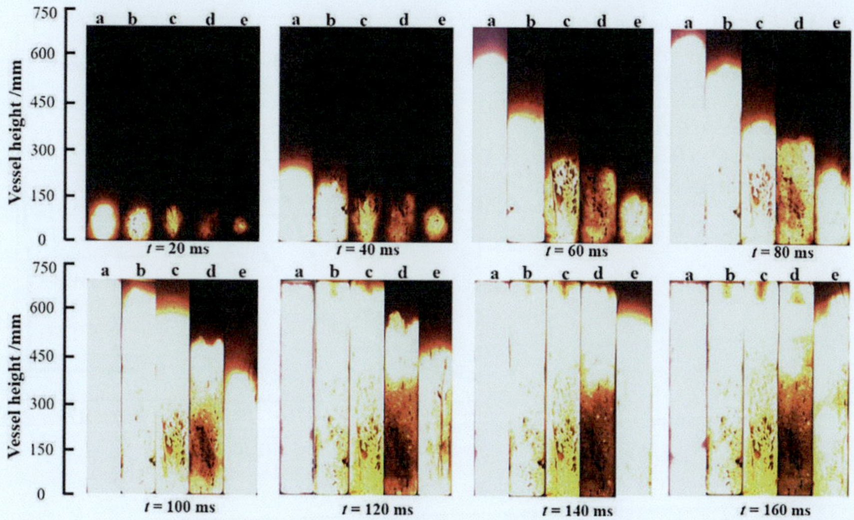

Fig. 3.5 Effect of mist concentration on the flame propagation process of methane and coal dust hybrid explosions ($c = 8\%$; $d_{50} = 76.6$ μm; $q = 300$ g/m^3; Q(a) = 0 g/m^3, Q(b) = 141 g/m^3, Q(c) = 282 g/m^3, Q(d) = 423 g/m^3, Q(e) = 564 g/m^3)

flame appeared an accelerated extinction phenomenon. Such as the height of flame front was decreased from 287 to 67 mm at 40 ms and from 704 to 247 mm at 80 ms respectively. And the effect of mist on the flame propagation of initial stage was more dramatically. The change history of flame front height with time under different mist concentrations could be obtained according to the Fig. 3.5, as shown in Fig. 3.6. The history slope was obviously decreased with increasing mist concentration. And the time for the flame reached the vessel top end was obviously delayed (from 84.0 to 186.5 ms). It indicates that the reaction rate of methane and coal dust hybrid explosions was obviously reduced after adding mist and the inhibition effect was gradually increased with increasing mist concentration.

The flame propagation velocity can be obtained by deriving the change history of flame front height with time. Figure 3.7a presents the effect of mist concentration on the flame propagation velocity of methane and coal dust hybrid explosions ($c = 8\%$; $d_{50} = 76.6$ μm; $q = 300$ g/m^3). The flame propagation velocity history was obviously decreased with increasing mist concentration. It was mainly reflected in the decrease of velocity history slope, the reduction of $v_{\max}$ (from 17.7 to 7.6 m/s) and the delay of its appearance moment (from 44.0 to 110.5 ms). And the $v_{\max}$ was decreased by 57.06%. Meanwhile, the time for the flame to reach the top end of vessel was also delayed by 102.5 ms. It indicates that the inhibition effect of ultrafine water mist on the hybrid explosions flame propagation was obviously enhanced with increasing mist concentration. Figure 3.7b, c shows the effects of coal dust concentration and diameter on the flame propagation velocity of hybrid explosions under the mist condition. It can be seen that the flame propagation velocity presented a trend of rising firstly and then decreasing with increasing coal dust concentration.

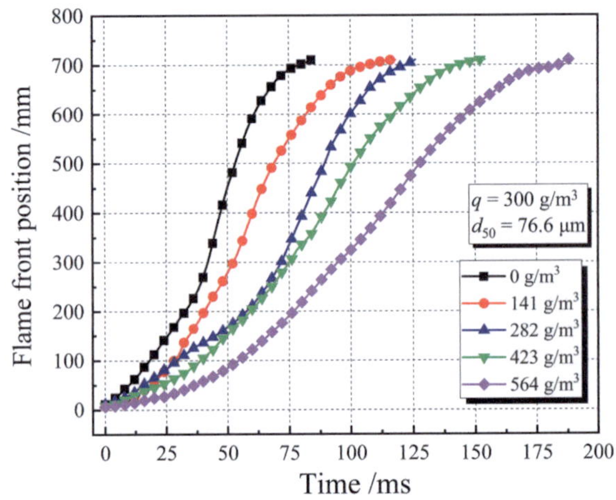

Fig. 3.6 Flame front height of methane and coal dust hybrid explosions under different mist concentrations ($c = 8\%$; $d_{50} = 76.6$ μm; $q = 300$ g/m^3; $Q = 0$, 141, 282, 423 and 564 g/m^3)

In particular, v_{max} was first increased from 6.9 to 12.8 m/s and then decreased to 8.0 m/s. As $q = 300$ g/m^3, the flame propagation velocity was the fastest ($v_{max} = 12.8$ m/s). The slope of velocity history and the time of flame propagation also presented the similar change trend. As the coal dust diameter increased, the flame propagation velocity was decreased successively. v_{max} was reduced from 48.7 to 10.0 m/s and its appearance moment was prolonged from 48.0 to 90.6 ms. It indicates that the concentration and diameter of coal dust could obviously affect the inhibition effect of ultrafine water mist on the gas and dust hybrid explosions.

3.2.3 *Analysis of Explosion Pressure*

Figure 3.8a–c shows the effects of mist concentration, coal dust concentration and diameter on the pressure history of methane and coal dust hybrid explosions. The pressure history was decreased successively with the increase of mist concentration ($c = 8\%$; $d_{50} = 76.6$ μm; $q = 300$ g/m^3), as shown in Fig. 3.8a. It was mainly reflected in the delay of time for pressure history began to rise (from 30.5 to 95.2 ms). ΔP_{max} was obviously decreased from 0.58 to 0.47 MPa and its appearance moment was delayed from 157.5 to 299.2 ms. It was decreased by 18.97%. This indicates that the intensity of hybrid explosions could be effectively reduced after adding mist, and the reduction extent was increased with increasing mist concentration. In Feng's experiment [5], the explosion pressure of 6% methane concentration and coal dust with 65 μm median diameter under 300 g/m^3 coal dust concentration was 0.61 MPa. Compared with our experiment, ΔP_{max} was increased by 4.92% although the methane concentration was slightly lower under no mist condition. It was because that the median diameter of coal dust in this experiment was slightly larger than that of the Feng's experiment. And the effect of mist concentration on ΔP_{max} also presented the similar change trend. Figure 3.8b, c present the effects of coal dust concentration and

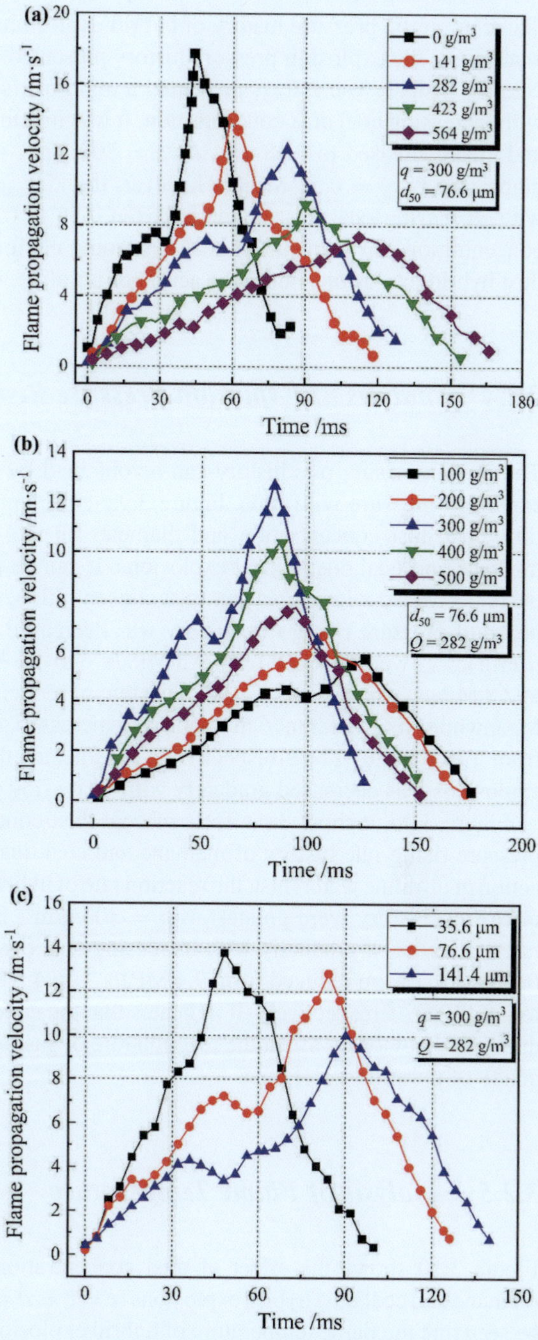

Fig. 3.7 **a** Effect of mist concentration on the flame propagation velocity of methane and coal dust hybrid explosions ($c = 8\%$; $d_{50} = 76.6$ μm; $q = 300$ g/m^3; $Q = 0$, 141, 282, 423 and 564 g/m^3). **b** Effect of coal dust concentration on the flame propagation velocity of methane and coal dust hybrid explosions ($c = 8\%$; $d_{50} = 76.6$ μm; $Q = 282$ g/m^3; $q = 100$, 200, 300, 400 and 500 g/m^3). **c** Effect of coal dust diameter on the flame propagation velocity of methane and coal dust hybrid explosions ($c = 8\%$; $Q = 282$ g/m^3; $q = 300$ g/m^3; $d_{50} = 35.6$, 76.6 and 141.4 μm)

diameter on the pressure history of hybrid explosions. Under the action of ultrafine water mist, the explosion pressure history presented the similar change trend to the velocity history. The ΔP_{max} presented a trend of rising firstly and then decreasing with increasing coal dust concentration. It was first increased from 0.40 to 0.52 MPa and then decreased to 0.48 m/s. As $q = 300$ g/m^3, the explosion pressure was the highest ($\Delta P_{max} = 0.52$ MPa). However, the ΔP_{max} presented a decreasing trend with increasing coal dust diameter (from 0.56 to 0.48 MPa). It indicates that the concentration and diameter of coal dust could obviously affect the intensity of gas/dust hybrid explosions under the action of ultrafine water mist.

3.2.4 Analysis of Explosion Pressure Rising Rate

The pressure rising rate history can be obtained by deriving the change history of explosion pressure with time. Figure 3.9a–c present the effects of mist concentration, coal dust concentration and diameter on the pressure rising rate history of methane and coal dust hybrid explosions. It can be seen that the history was obviously decreased with increasing mist concentration, as shown in Fig. 3.9a. The two peaks of pressure rising rate history was decreased obviously. The $(dP/dt)_{first}$ and $(dP/dt)_{second}$ were reduced from 9.3 MPa/s^{-1} to 4.1 MPa/s^{-1} and from 7.2 MPa/s^{-1} to 2.8 MPa/s^{-1} respectively. And the extent of reduction reached 55.91 and 61.11%. Meanwhile, its appearance moment was increased from 78.8 ms to 141.4 ms and from 126.4 ms to 224.8 respectively. It indicates that the reaction rate of hybrid explosions was decreased gradually with increasing mist concentration.

Figure 3.9b, c show the effects of coal dust concentration and diameter on the pressure rising rate history of methane and coal dust hybrid explosions. Under the action of ultrafine water mist, the reaction rate of hybrid explosions and the peak value of (dP/dt) history were greatest as $q = 300$ g/m^3. Besides, the explosion reaction rate was reduced gradually with increasing coal dust diameter. The $(dP/dt)_{first}$ and $(dP/dt)_{second}$ were reduced from 9.4 MPa/s^{-1} to 4.8 MPa/s^{-1} and from 7.5 MPa/s^{-1} to 2.9 MPa/s^{-1} respectively. It indicates that the concentration and diameter of coal dust could obviously affect the reaction rate of gas/dust hybrid explosions under the action of ultrafine water mist.

3.2.5 Analysis of Flame Temperature

Figure 3.10 shows the effect of mist concentration on the flame temperature of methane and coal dust hybrid explosions ($c = 8\%$; $d = 76.6$ μm; $q = 300$ g/m^3). It can be seen that the flame temperature of hybrid explosions could be effectively reduced after adding mist and the reduction extent was increased gradually with the increase of mist concentration. It was mainly reflected in the decreases of temperature history slope and T_{max}. T_{max} was decreased from 1646 to 817 °C and its appearance moment

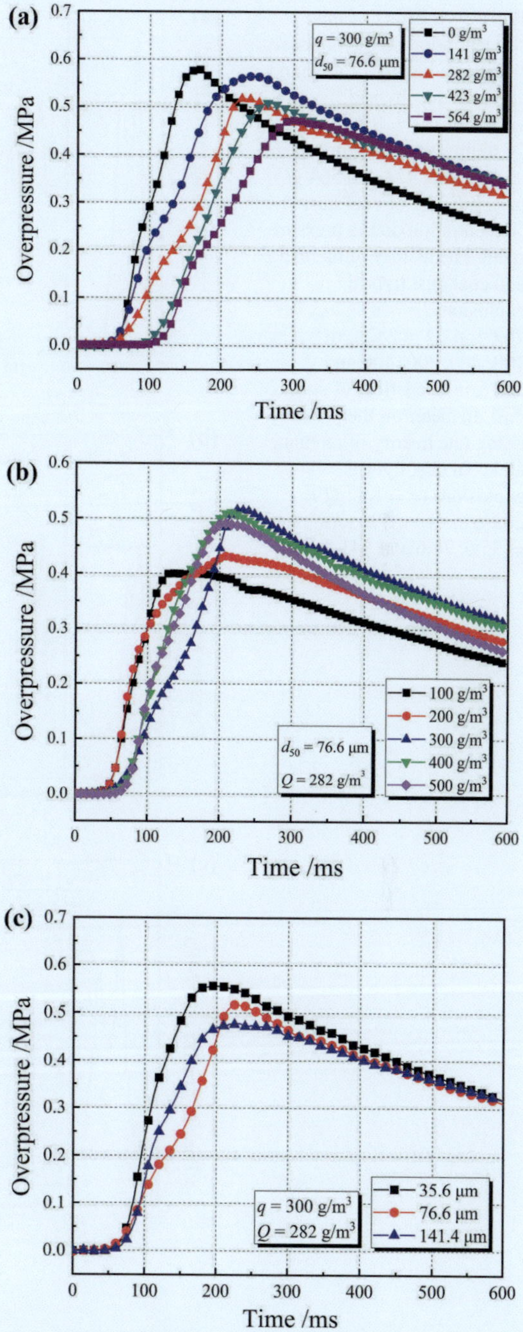

Fig. 3.8 **a** Effect of mist concentration on the pressure history of methane and coal dust hybrid explosions ($c = 8\%$; $d_{50} = 76.6$ μm; $q = 300$ g/m^3; $Q = 0$, 141, 282, 423 and 564 g/m^3). **b** Effect of coal dust concentration on the pressure history of methane and coal dust hybrid explosions ($c = 8\%$; $d_{50} = 76.6$ μm; $Q = 282$ g/m^3; $q = 100$, 200, 300, 400 and 500 g/m^3). **c** Effect of coal dust diameter on the pressure history of methane and coal dust hybrid explosions ($c = 8\%$; $Q = 282$ g/m^3; $q = 300$ g/m^3; $d_{50} = 35.6$, 76.6 and 141.4 μm)

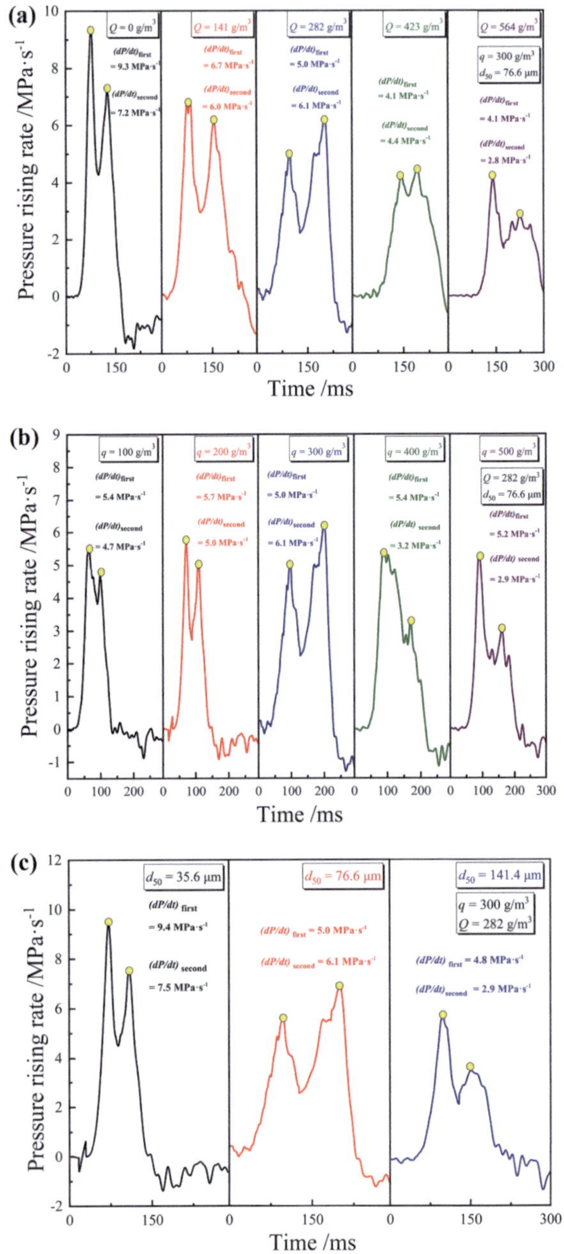

Fig. 3.9 **a** Effect of mist concentration on the pressure rising rate history of methane and coal dust hybrid explosions ($c = 8\%$; $d_{50} = 76.6$ μm; $q = 300$ g/m^3; $Q = 0$, 141, 282, 423 and 564 g/m^3). **b** Effect of coal dust concentration on the pressure rising rate history of methane and coal dust hybrid explosions ($c = 8\%$; $d_{50} = 76.6$ μm; $Q = 282$ g/m^3; $q = 100$, 200, 300, 400 and 500 g/m^3). **c** Effect of coal dust diameter on the pressure rising rate history of methane and coal dust hybrid explosions ($c = 8\%$; $Q = 282$ g/m^3; $q = 300$ g/m^3; $d_{50} = 35.6$, 76.6 and 141.4 μm)

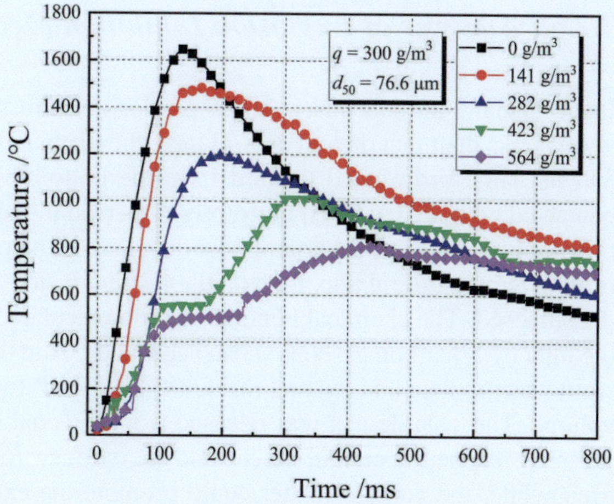

Fig. 3.10 Effect of mist concentration on the flame temperature of methane and coal dust hybrid explosions ($c = 8\%$; $d = 76.6$ μm; $q = 300$ g/m^3; Q(a) = 0 g/m^3, Q(b) = 141 g/m^3, Q(c) = 282 g/m^3, Q(d) = 423 g/m^3, Q(e) = 564 g/m^3)

was delayed from 142 to 450 ms. It was decreased by 50.18%. This indicates that the ultrafine water mist had a strong endothermic effect on the hybrid explosion flame and the effect could be enhanced with increasing mist concentration.

Figure 3.11 shows the change law of T_{max} with the increases of coal dust concentration and diameter under the action of ultrafine water mist. T_{max} also presented a trend of rising firstly and then decreasing with increasing coal dust concentration. It was first increased from 1018 to 1228 °C and then decreased to 1122 °C under the 35.6 μm coal dust diameter. As $q = 300$ g/m^3, T_{max} reached 1228 °C. However, a decrease trend was presented with the coal dust diameter increased. It was decreased from 1228 to 1111 °C under the 300 g/m^3 coal concentration. It indicates that the concentration and diameter of coal dust could obviously affect the flame temperature of gas/dust hybrid explosions under the action of ultrafine water mist.

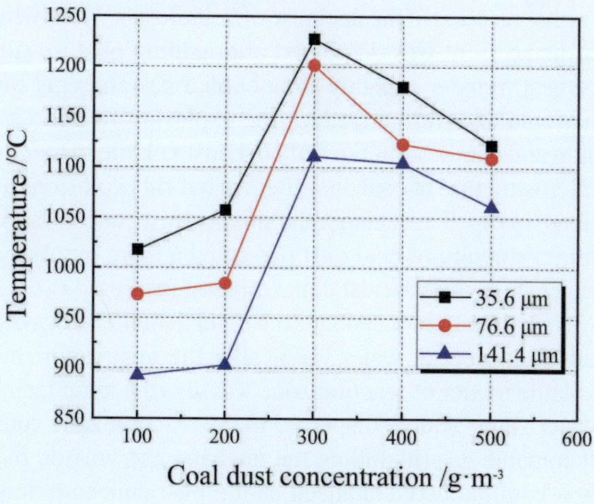

Fig. 3.11 Effects of coal dust concentration and diameter on the flame temperature of methane and coal dust hybrid explosions

3.2.6 Analysis of Explosion Inhibition Mechanism

In the methane and coal dust hybrid explosions system, the ignition energy of methane was far less than that of coal dust. Hence, the methane was ignited prior to coal dust and its flame propagated upward from the bottom end of vessel. And the flame contacted with the coal dust uniformly dispersed in the vessel and provided heat for the pyrolysis and combustion of coal dust during the propagation. Under the action of high temperature flame, the volatile matters inside the coal dust were constantly decomposed. The chemical bond of macromolecular structure was broken, forming the light hydrocarbon gaseous substance and tar. And the precipitated volatile matter would accumulate and expand inside the coal dust, further increasing the coal dust volume. The volatile gas was released from the coal dust surface after its internal pressure reached a certain extent and the surface formed the pore structure [6]. Meanwhile, the volatile matter could promote the explosion reaction and transfer heat to coal dust, resulting in more violent pyrolysis reaction and volatile matter release [7]. Besides, the remaining coke would happen the combustion reaction under the action of high temperature flame after the volatile was separated from coal dust. The duration of coke combustion was longer compared with the methane explosion, resulting in the higher explosion intensity inside the vessel during the gas/dust hybrid explosions.

Figure 3.12 shows the interaction process of gas/liquid/solid three phases during methane and coal dust hybrid explosions under the action of ultrafine water mist. The inhibition of hybrid explosions by the mist was mainly attributed to the combination of physical and chemical effects. The physical effect mainly included the heat absorbing, heat radiation blocking, gas concentration diluting [8]. The mist with 10 μm diameter had a large specific surface area, which could completely vaporize and quickly absorb heat after contacting with the flame front. Thus, the temperature of flame front was decreased and the reaction rate was slowed down. And the pyrolysis and release rates of volatile matter inside the coal dust were also reduced, which further resulted in the decrease of volatile gas. Meanwhile, the heat radiation intensity of flame front was decreased after adding mist, resulting in the obvious reduction of heat transfer capacity to unburned gas and coal dust in the reaction zone. And it would also further reduce the explosion reaction rate of methane and the emission and combustion rates of coal dust volatile matter. It can be seen from the surface microstructure of coal dust after the hybrid explosions under different mist concentrations (in Fig. 3.12). Under the smaller mist concentration ($Q = 141$ g/m^3), the surface microstructure of coal dust presented a more pyrolysis holes and a greater combustion extent. As the mist concentration increased ($Q = 423$ g/m^3), the pore structure was obviously reduced and the combustion extent was also decreased. Besides, the dilution effect of water vapor after the mist vaporization on the premixed gas and volatile matter of reaction zone was also the main factor of explosion inhibition. The water vapor widely dispersed in the reaction zone could dilute the concentration of flammable gas (including the methane and volatile matter). And the dilution effect was continuously enhanced as the mist concentration increased. Furthermore, the

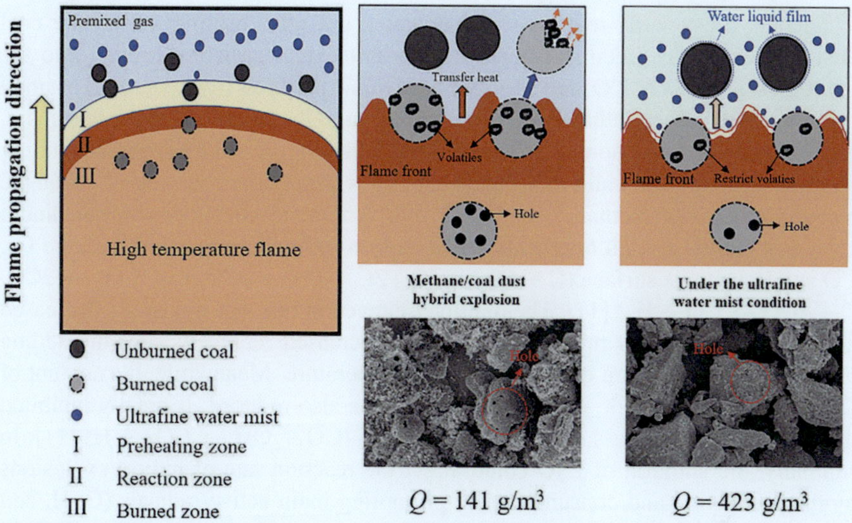

Fig. 3.12 Schematic diagram of methane and coal dust hybrid explosions inhibition by the ultrafine water mist

water vapor could also reduce the oxygen concentration in the area near the coke, reducing the combustion rate of coal dust. And the mist could adsorb in the pores of coal dust and form a water liquid film on its surface, further blocking the reaction of coke and oxygen. Macroscopically, the flame propagation velocity and explosion pressure was decreased significantly with the increase of mist concentration. Especially the decrease of flame temperature.

The mist could also give play to obvious chemical inhibition on the hybrid explosions. First, the water vapor was widely distributed in the reaction zone during the methane explosion, and as a third body would participate in the chain reaction. This process was mainly reflected in the following aspects. Amount of high-energy radicals during explosion reaction would collide with the water molecules. As a result, the radicals would either become inactive with the energy, which were used to activate the branch chain reaction, absorbed by water molecules or react with the water molecules which interrupted the branch chain reaction. It would greatly weaken the reaction rate of methane explosion, resulting in the obvious decrease of flame propagation velocity. Meanwhile, water molecules would also participate in the chain reaction and interfere with the main reaction path. Especially the effect of water molecules on the rates of promoting main active radicals (O, H, and OH) production and consumption during the methane explosion. Namely, the key reaction steps of producing the main active radicals were prevented ($HO_2 + CH_3 \rightarrow OH + CH_3O$, $H + O_2 \rightarrow O + OH$) and the reaction steps of consuming the main active radicals were promoted ($2H + H_2O \rightarrow H_2 + H_2O$, $H + OH + M \rightarrow H_2O(+M)$). Besides, the reaction rate of promoting methane production ($C_2H_6(+M) \rightarrow 2CH_3(+M)$) was greater than that of promoting methane consumption ($CH_3 + O_2 \rightarrow O + CH_3O$) [9].

Second, the volatile matters were separated out after heating during the coal dust explosion. The chemical bond of macromolecular structure broken into the gaseous (N_2, O_2, CO_2, CO, and CH_4, etc.) and liquid products (hydrocarbon) of small molecules [10]. The methane was produced through the pyrolysis of fat side chain, and the carbon monoxide was produced through the pyrolysis of oxygen-containing functional group. The water molecule would show the similar chemical effect on the reaction process of methane volatilized from coal dust. After then, the remaining coke was melted at high temperature and happen multiphase reaction with the O_2, CO and H_2O on its surface ($C + O_2 \rightarrow CO_2$, $2C + O_2 \rightarrow 2CO$, $C + CO_2 \rightarrow 2CO$, $C + H_2O \rightarrow CO + H_2$) [11]. The multiphase reaction rate was obviously decreased and the amount of CO production was also decreased relatively after the adding mist due to the reduction of carbon surface temperature. Meanwhile, the amount of OH produced by volatile matter combustion was also reduced. It greatly inhibited the main reaction process of coal dust explosion ($CO + OH \rightarrow CO_2 + H$) [11]. In summary, the addition of H_2O could affect the reaction rate of hybrid explosions by the production and consumption of promoting main active radicals (O, H, and OH). Macroscopically, the reaction rate of methane and coal dust hybrid explosions was decreased. Especially the pore structure of coal dust surface was obviously reduced and the combustion extent was also decreased. And the weakening effect was enhanced with the increase of mist concentration.

3.3 Conclusions

The effect of ultrafine water mist on the flame propagation characteristics and intensity of gas/dust hybrid explosions were carried out by the experiment. Combined with the change of explosion parameter under different mist conditions, the corresponding relationship between flame propagation and pressure rise of gas/dust hybrid explosions was clarified. And the effects of mist and coal dust parameters on the hybrid explosions and the inhibition mechanism were researched in depth. Conclusions are as follows:

First, the flame propagation and pressure rise of gas/dust hybrid explosions was corresponding. However, the corresponding relationship was not concerned in current literature. The flame front appeared an obvious thermal diffusion instability and presented an irregular structure due to the endothermic effect of the mist. The gas/dust hybrid explosions could be obviously inhibited by the mist. As the mist concentration increased, the flame gradually changed from the bright white to dark red. v_{max} was decreased by 57.06% and its appearance moment was delayed. It indicates that the mist had a strong endothermic effect on the explosion flame. The coal dust concentration and diameter could also obviously affect the inhibition of mist on the flame propagation. The v_{max} presented a trend of rising firstly and then decreasing with increasing coal dust concentration. However, a decrease trend was presented with increasing coal dust diameter.

Second, the pressure history of hybrid explosions inside the closed vessel appeared two acceleration rise processes and the (dP/dt) history showed two peaks. With the mist concentration increased, the time for the pressure history began to rise was delayed. The ΔP_{max} was decreased by 18.97% and its appearance moment was prolonged. And the $(dP/dt)_{first}$ and $(dP/dt)_{second}$ were reduced by 55.91% and 61.11% respectively. Besides, ΔP_{max} and (dP/dt) also presented the similar change trend to v_{max} with the changes of coal dust concentration and diameter.

Third, the flame temperature of hybrid explosions could be effectively reduced after adding mist and T_{max} was decreased by 50.18% with increasing mist concentration. T_{max} presented a trend of rising firstly and then decreasing with increasing coal dust concentration under the action of mist. However, a decreasing trend was presented with the coal dust diameter increased. The inhibition mechanism of mist on gas/dust hybrid explosions was mainly attributed to the combination of physical and chemical effects.

For effective inhibition of gas/dust hybrid explosions with ultrafine water mist in petrochemical enterprises safe protection, four suggestions are proposed based on present researches: (1) Enterprises involving flammable gas should avoid the existence of dust with the smaller particle size and away from the dust concentration of maximum explosion intensity as far as possible. (2) To further improve the inhibition effect of mist on gas/dust hybrid explosions, additives that can enhance the heat absorption capacity and chemical effect should be added. (3) The spray equipment that can achieve the large atomization quantity with the rapid response should be installed in the area with the high frequent hybrid explosions. (4) The flame propagation and pressure rise presented a corresponding relationship. Hence, the distribution areas of personnel and equipment should avoid flame propagation as much as possible to reduce the injury caused by explosion overpressure. Meanwhile, the mist concentration for realizing the complete inhibition of gas/dust hybrid explosions should be determined.

References

1. Cao XY, Wei HY, Wang ZR, Wang YN, Fan LT, Lu YW. Effect of obstacle on the H_2/CO/Air explosion characteristics under lean-fuel conditions. Fuel 2022; 319: 123834.
2. Cao XY, Ren JJ, Bi MS, Zhou YH, Wang QJ. Experimental research on methane/air explosion inhibition using ultrafine water mist containing additive. J Loss Prev Process Ind 2016; 43: 352–360.
3. Zhang PP, Zhou YH, Cao XY, Gao XL, Bi MS. Mitigation of methane/air explosion in a closed vessel by ultrafine water fog. Saf Sci 2014; 6: 1–7.
4. Jing Q, Wang D, Liu QM, Chen X, Shen Y, Wang ZS, Zhong YP. Inhibition effect and mechanism of ultrafine water mist on CH_4/air detonation: quantitative research based on CFD technology. Process Saf Environ Prot 2021; 148: 7592.
5. Feng X, Ren JJ, Pu MZ, Chen B, Bi MS. Suppression effect of ultrafine water mist on methane-coal dust hybrid explosions. Powder Technol 2022; 406: 117590.
6. Nie BS, Liu XF, Yang LL, Meng JQ, Li XC. Pore structure characterization of different rank coals using gas adsorption and scanning electron microscopy. Fuel 2015; 158: 908–917.

7. Zhou YH, Gan B, Jiang HP, Huang L, Gao W. Investigations on the flame propagation characteristics in methane and coal dust hybrid explosions. Shock waves 2022; 42 (01): 167–175.
8. Cao XY, Ren JJ, Zhou YH, Wang QJ, Gao XL, Bi MS. Suppression of methane/air explosion by ultrafine water mist containing sodium chloride additive. J Hazard Mater 2015; 285: 311–318.
9. Liang YT. Kinetic characteristics and inhibition mechanism of gas explosion reaction. First ed. Beijing: Science Press; 2013.
10. Liu ZT, Zhang SS, Xi RZ, Guo RL, Liu BB, Lin S. Analysis on residual gas characteristics of coal dust explosion in confined space. J Coal Sci 2015; 40(07): 1574–1579.
11. Yang LL, Reaction dynamic characteristics of coal dust and gas explosion and disaster-causing mechanism, in: China Univ. Min. Technol, Beijing, 2018. (in Chinese).

Chapter 4
Research on the Inhibition Characteristics of Ultrafine Water Mist on Gas/Dust Two-Phase Mixture Explosions

4.1 Experimental Apparatus and Procedures

Figure 4.1 provides a structural diagram of experiment apparatus for inhibiting gas/dust two-phase explosions using the mist. It mainly included eight subsystems, and the main structure of experimental apparatus had been introduced in the previous work [1]. The gas supply system consisted of a gas cylinder and a premixed tank. The premixed gas (composed of methane and air) with determined volume concentration was configured in the premixed tank. The dust spraying system consisted of an electromagnetic valve, a one-way valve and a Hartmann dust spraying device. The Hartmann dust spraying device was installed on the flange of explosion vessel bottom end and connected to the premixed tank by a tube bundle. The closing and opening of the solenoid valve was control through the self-programming. The coal dust parameter was determined and evenly placed in the Hartmann dust groove. As the high-speed premixed gas sprayed from the bottom tube bundle flowed through the Hartmann dust spraying device, a bottom-up vortex was formed due to the obstruction of the hemispherical "mushroom cap", further spraying the coal dust and forming a dust cloud. This type of dust spraying device could realize relatively uniform dust cloud. The ultrasonic atomization device was adopted to produce the mist and it was installed at the vessel top end. By measuring, the mist diameter (SMD) was 10.03 μm and it was suspended inside the vessel due to the action of drag. The atomization rate of mist was tested by a high-precision electronic balance (Q = 70.05 g/m^3). According to the atomization rate and time, four types of water mist concentrations can be achieved inside the vessel.

The gas/dust two-phase mixture was ignited by two electrodes with 8 kV high-voltage. The pressure history was collected using a high-frequency dynamic pressure transmitter (range: 0–2.5 MPa; response frequency: 200 kHz). The temperature change of explosion flame under the mist condition was recorded using an S-type platinum rhodium thermocouple (range: 0–1600 °C). A high-speed camera

Z. Wang and X. Cao, *Mechanism of Composite Explosion Inhibition and Venting Protections*, https://doi.org/10.1007/978-981-95-4809-5_4

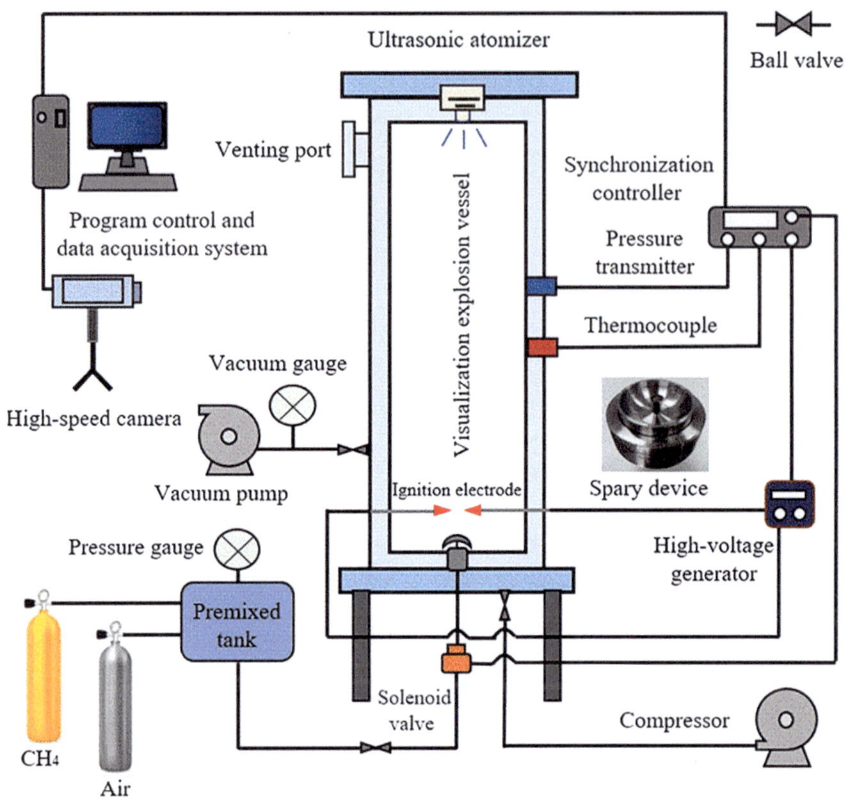

Fig. 4.1 Schematic diagram of experiment apparatus for inhibiting methane and coal dust mixture explosions by the mist

(V1212) was used to gather the changes in flame morphology (2000 fps). An eight-channel high frequency data acquisition card (USB2871; the acquisition frequency: 200 kHz) was adopted to realize the program control and data acquisition. Through the programming language, the orders of the image acquisition, the powder spraying, the explosion pressure and temperature data acquisitions, the ignition were achieved and their adjacent interval time was determined [2].

The explosive limit of methane could be evidently affected after adding coal dust, Hence, to achieve effective comparative analysis, four methane concentrations ($c = 4$, 6, 8 and 10%) were adopted during the explosion inhibition. By calculation, a mixed gas with determined pressure and concentration was configured inside the premixed tank. Meanwhile, the coal dust could be raised by the premixed gas and realize the determined methane concentration inside the vessel. And the pressure of vessel interior was 0.1 MPa before igniting. By a vacuum pump, the vessel interior reached a predetermined vacuum degree before dust injection. After the solenoid valve was opened, the coal dust was carried by the premixed gas and evenly dispersed inside the vessel. The explosion intensity was the greatest under 0.50 MPa injection pressure

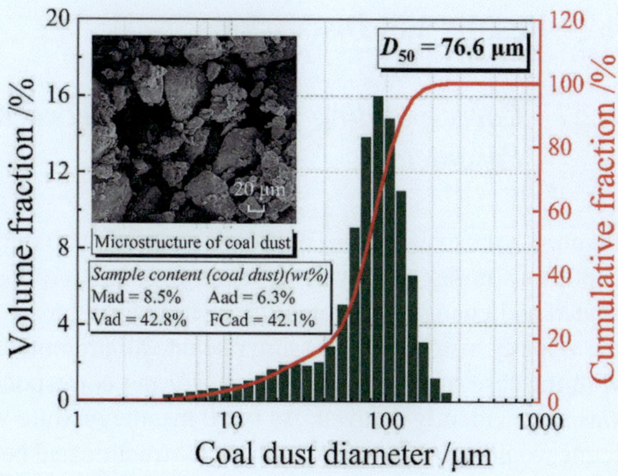

Fig. 4.2 The test data of coal dust physical parameter

(P_{st}) and 700 ms injection time and 100 ms settling time by repeated experiments, indicating that the coal dust distribution was the most uniform inside the vessel. In addition, the determined mist concentration was achieved before ignition. Four kinds of mist concentrations were adopted during explosion inhibition ($Q = 0$, 141, 282 and 423 g/m^3). After the mixture was ignited, the flame rapidly spread and developed upwards along the axial direction. The coal dust with determined diameter ($D_{50} =$ 76.6 μm) and concentration ($q = 300$ g/m^3) after injection was placed in the groove. Its component was tested and the morphological structure (×2000) was analyzed before the experiment as can be seen in Fig. 4.2.

The component content of coal dust was measured by the industrial analysis (M_{ad} = 8.5%; A_{ad} = 6.3; V_{ad} = 42.8%; FC_{ad} = 42.0%). Besides, its diameter distribution was also tested by a laser particle size analyzer. The morphology showed a uniform distribution state and smooth surface structure, and the diameter history also presented a normal distribution. Simultaneously, the pyrolysis characteristics of coal dust under the high temperature with a heating rate of 10 K/min was tested. It underwent three stages: the heat absorption, weight loss and stability. Before 295 °C, the coal dust exhibited an endothermic process. Thereafter, the pyrolysis occurred and the weight loss rate was gradually increased. At 578 °C, the weight loss rate was the fastest and its corresponding DTG history reached the peak value. Subsequently, the weight loss rate was evidently reduced. After reaching 710 °C, the weight loss of coal dust reached a stable state. The functional group changes of coal dust before and after explosion under the mist inhibition were tested by an infrared spectrometer. After ignition, the flame of mixture explosions developed as a spherical structure and accelerated upward propagation along the pipeline. During the propagation, the flame and mist realized the heat-mass transfer. The deviation value of ΔP_{max} was less than 5.0% under the mist condition through the repeated experiments, indicating the good repeatability and reliability during experiment.

4.2 Results and Discussions

4.2.1 Corresponding Relationship Between Explosion Parameters

Figure 4.3 presents the rise of pressure history with the flame propagation of mixture explosions under the action of mist ($Q = 282$ g/m^3; $c = 4\%$). As the flame propagated and changed, the pressure rise showed the same correspondence. However, the velocity history and pressure rise rate all presented a decreasing trend compared with the case no mist. The moment for the corresponding relationship appearance was also evidently delayed. As $t = 0$ ms, the mixture was ignited. Nevertheless, the flame would not develop in a spherical structure and become dark-irregular structure ($t = 20$–40 ms). Constrained by the wall surface, the flame accelerated upward propagation ($t = 40$–60 ms). But the flame front still showed a dark-irregular structure and the moment for the pressure begin to rise (t_{rise}) was evidently prolonged ($t =$ 68.9 ms) [3]. At 112.0 ms, the velocity peak appeared ($v_{\max} = 11.1$ m/s). And it was evidently decreased and the corresponding moment was also prolonged compared with the case no mist. The reduction in flame front area was evidently increased as it continued to propagated, further leading to the obvious reduction of velocity history. But it still maintained a relatively high value. During this stage, the pressure history experienced the first acceleration rise process as a result of the flame acceleration propagation ($t = 68.9$–160.0 ms). However, the first peak of (dP/dt) history ($(dP/dt)_{\text{first}} = 5.0$ MPa/s^{-1}) was evidently reduced and the corresponding moment ($t =$ 135.5 ms) was delayed compared with the case no mist. It explains that the reaction rates of the mixture explosions were greatly weakened after adding mist, thereby reducing its intensity. Due to the reflection of pressure waves generated by the vessel top end, the ability of flame propagation was greatly weakened [1]. Nevertheless, the pressure rise again caused by the increase in temperature due to the rapid accumulation of heat inside the vessel as the explosion reaction progressed ($t = 160.0$–244.5 ms). But the second peak of (dP/dt) history ($(dP/dt)_{\text{second}} = 5.9$ MPa/s^{-1}) was also evidently reduced and corresponding moment was delayed ($t = 190.0$ ms) compared with the case no mist. At this moment, the flame was the brightest and the reaction was the most intense. Thereafter, the brightness started to decrease and the rise rate of pressure was quickly decreased. At 300.0 ms, $\Delta P_{\max}$ appeared. Subsequently, the pressure was dramatically reduced as the accelerated extinguishing of explosion flame.

4.2.2 Comparison of Flame Structure

Figure 4.4 shows the comparison of velocity histories at four methane concentrations under action of water mist ($Q = 282$ g/m^3). Compared with the case no mist, the brightness of flame was evidently reduced during the initial stage and it changed

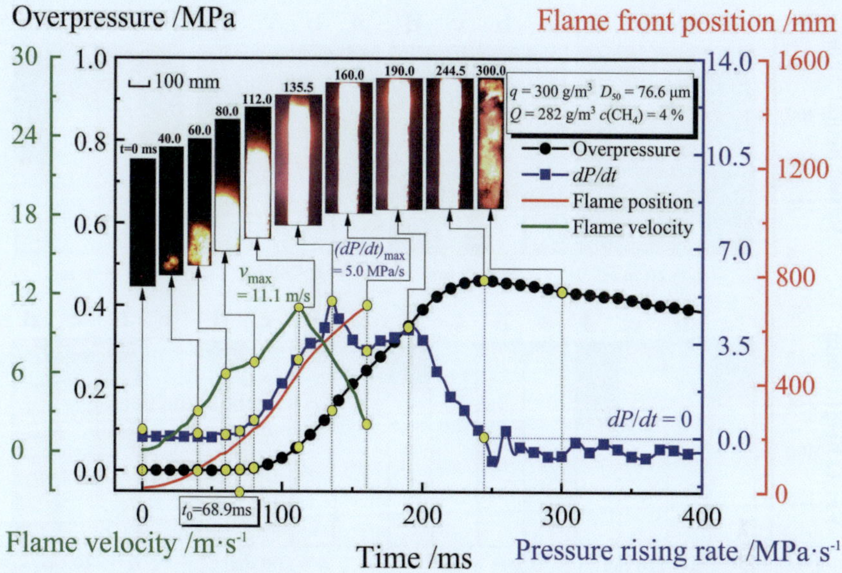

Fig. 4.3 Corresponding relationship between explosion parameters under the action of mist ($Q =$ 282 g/m^3; $c = 4\%$)

from the bright-white to dark-red due the endothermic effect of mist. Meanwhile, its front no longer exhibited a smooth structure and its wrinkle degree was evidently enhanced. It fully demonstrates that the flame happened an evident thermal-diffusion instability due to the heat absorption action of mist, leading to instability of the flame. Simultaneously, the time of flame propagation inside the vessel experienced a trend of firstly decreasing and then increasing as the methane concentration raised. At $c = 6\%$, the flame first propagated to the vessel top end, and follow by 8% methane concentration. As $c = 4\%$ and 10%, the time for reaching the vessel top end was evidently prolonged. And the corresponding flame exhibited a more unstable structure. After adding mist, the times for mixture explosion flames with different methane concentrations to reach the vessel top end all were evidently prolonged. And the weakening effect of mist on the weaker flame was more evident. Besides, the velocity history also presented the same trend of change as pressure as the methane concentration raised under the action of mist (in Fig. 4.5). It was clearly reflected by the changes of velocity history slope and its peak value. As $c = 6\%$, the explosion reaction rate was the fastest, leading to the highest velocity history. Follow by 8% methane concentration. The velocity history was the smallest as $c = 10\%$. This indicates that methane concentration had a significant impact on the propagation of mixture explosion flames under the mist inhibition, especially for the weaker explosion flame.

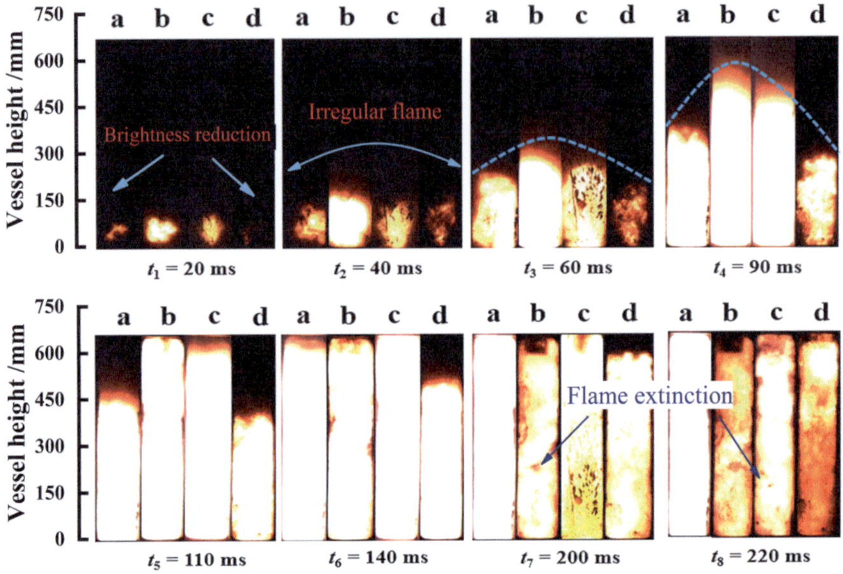

Fig. 4.4 Comparisons of flame propagation processes at four methane concentrations under action of water mist ($Q = 282$ g/m^3)

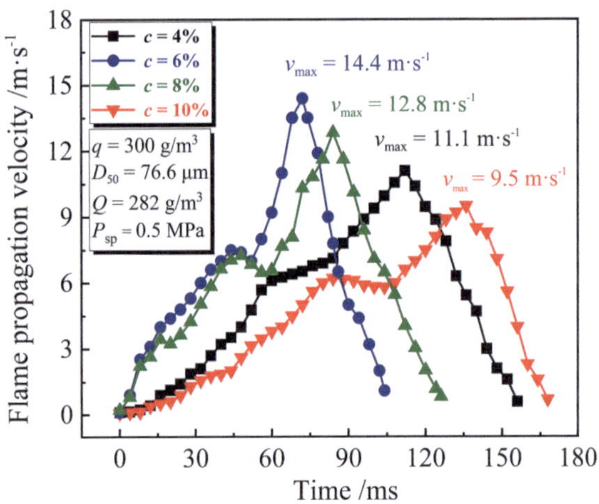

Fig. 4.5 Comparisons of velocity histories at four methane concentrations under action of water mist ($Q = 282$ g/m^3)

4.2.3 Analysis of Explosion Pressure

Figure 4.6 shows the influence of methane concentration on the pressure histories without mist condition. Within the range of 4–10% methane concentration, ΔP_{max} and slope of pressure history were firstly increased and then reducing. As $c = 6\%$, the corresponding pressure was the greatest and its rising rate was the fastest. As

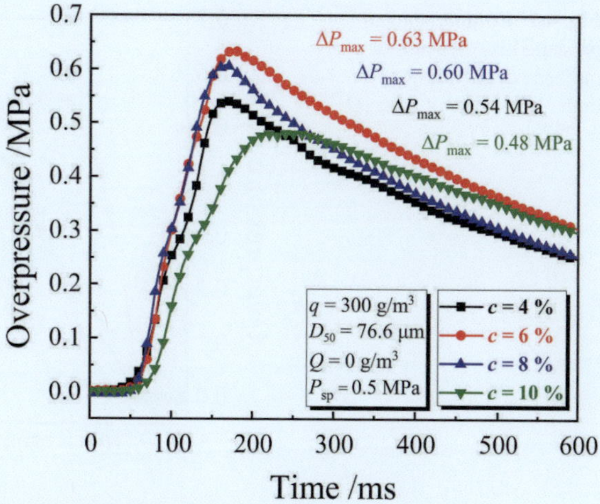

Fig. 4.6 Effect of methane concentration on the pressure histories without mist condition

$c = 10\%$, the pressure was the smallest and its rising rate was also the slowest. It explains that the methane concentration possessed a significant impact on the intensity of mixture explosions. Figure 4.7 presents the pressure histories of mixture explosions at four methane concentrations after adding mist ($Q = 282$ g/m^3). The pressure history emerged the similar change rule as the case no mist as $4 \leq c \leq 10\%$. However, the pressure all was obviously decreased and the t_{rise} was delayed under four different methane concentrations. As $c = 6\%$, ΔP_{max} was reduced from 0.63 to 0.56 MPa. Meanwhile, the t_{rise} was also prolonged from 44.2 to 70.8 ms. In particular, the decrease in pressure at a corresponding 10% methane concentration was more significant and the t_{rise} was obviously prolonged (from 61.4 to 109.9 ms). This shows that the mist could effectively weaken the intensity of gas/dust two-phase mixture explosions and its reaction rate. And the weakening effect was related to the methane concentration.

Figure 4.8 shows the ΔP_{max} and its decrease extent of mixture explosions with increasing methane concentration under four mist concentrations. It clear that ΔP_{max} at different methane concentrations was obviously decreased as the mist concentration increased (from 0 to 423 g/m^3). Meanwhile, the drop percentage of ΔP_{max} was also continuously improved as the mist concentration increased. The corresponding ΔP_{max} to 6 and 8% methane concentrations was reduced from 0.64 MPa and 0.61 MPa to 0.54 MPa and 0.51 MPa, respectively. The drop percentage reached 14.3 and 16.3%. As the change of methane concentration, the drop percentage of ΔP_{max} was gradually increased. In particular, the corresponding ΔP_{max} to 4 and 10% methane concentrations was reduced from 0.54 and 0.48 MPa to 0.43 and 0.37 MPa. And the corresponding drop percentage reached 20.9% and 22.8%, respectively. This shows that the weakening effect of mist on gas/dust mixture explosions for different methane concentrations had an obvious difference, and the weakening effect was more evident on the weaker explosion flame. At high temperature, the vapor pressure generated by mist gasification could evidently affect the internal pressure of closed vessel. The

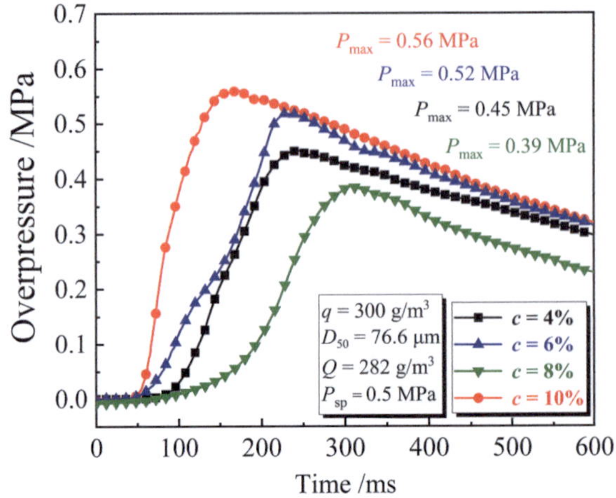

Fig. 4.7 Comparisons of pressure histories at four methane concentrations under the action of mist ($Q = 282$ g/cm^3)

main factors leading to the enhancement and inhibition of mixture explosions caused by mist were the combined effect of water mist cooling, which leads to a decrease in the pressure of gas phase components, and water mist vaporization, which leads to an increase in vapor pressure inside the vessel [4]. Meanwhile, both them could be evidently affected by the temperature rise rate and mist gasification rate. Hence, the pressure presented the evident difference with the methane and mist concentrations changed.

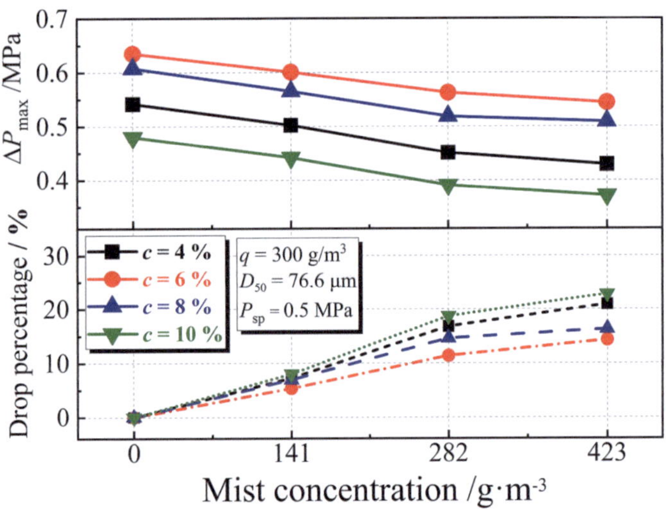

Fig. 4.8 ΔP_{max} and its drop percentage with the increases of methane and mist concentrations

4.2.4 Analysis of Explosion Pressure Rising Rate

Figure 4.9 illustrates the effect of methane concentration on the (dP/dt) histories without mist condition. As the methane concentration increased, $(dP/dt)_{first}$ and $(dP/dt)_{second}$ all appeared a trend of firstly rising and then reducing. Especially, the change in $(dP/dt)_{first}$ was more significant. As $c = 6\%$, $(dP/dt)_{first}$ was the greatest (10.8 MPa/s). As the methane concentration changed, it was obviously decreased. For 4% and 10% methane concentrations, $(dP/dt)_{first}$ was reduced to 8.3 and 7.0 MPa/s. It indicates that the change in methane concentration had a significant impact on the reaction rate of gas/dust two-phase mixture explosions. After adding mist, the reaction rate could be further affected. Figure 4.10 illustrates the influence of methane concentration on (dP/dt) history after adding mist ($Q = 282$ g/m^3). Compared with the condition no mist, the (dP/dt) of mixture explosions showed a decreasing change trend. As $c = 6\%$, the corresponding $(dP/dt)_{first}$ was decreased to 8.2 MPa/s and the decrease extent reached 24.1%. For 4 and 10% methane concentrations, $(dP/dt)_{first}$ was decreased to 5.0 and 3.9 MPa/s. And the corresponding decrease extent was 39.8 and 44.3%. This demonstrates that the explosion reaction rate could be evidently weakened by mist, and the weakening degree was related to the explosion intensity.

The $(dP/dt)_{first}$ was larger and its change was also more significant. Hence, it was considered as the $(dP/dt)_{max}$. Figure 4.11 illustrates $(dP/dt)_{max}$ and its decrease extent of mixture explosions with increasing methane and mist concentrations ($D_{50} = 76.6$ μm; $q = 300$ g/m^3). As the mist concentration increased, $(dP/dt)_{max}$ presented a decrease trend and its reduction extent was evidently increased. As $c = 6\%$, $(dP/dt)_{max}$ was the most greatest and corresponding reduction extent was the smallest.

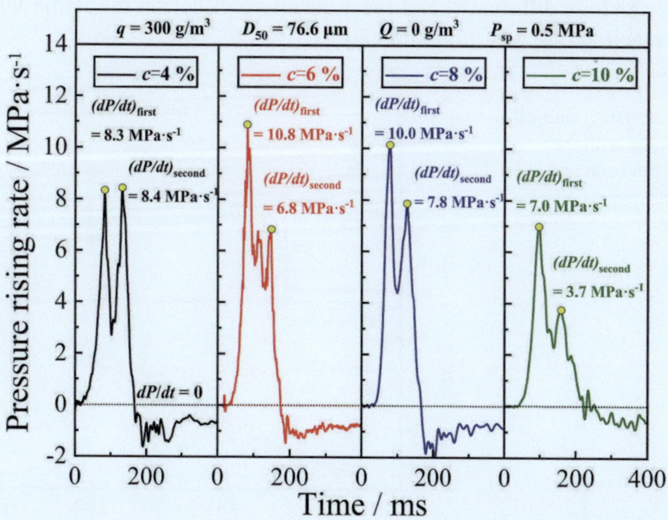

Fig. 4.9 Effect of methane concentration on (dP/dt) histories without mist condition

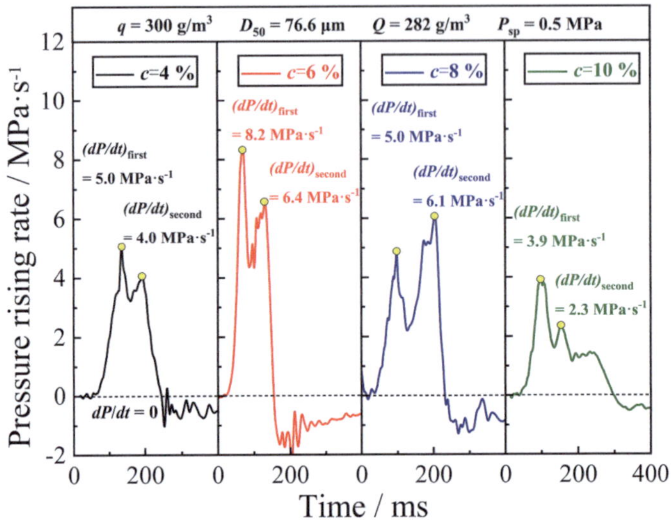

Fig. 4.10 Effect of methane concentration on (dP/dt) histories under the action of mist ($Q = 282$ g/m^3)

And it was reduced from 10.8 to 7.6 MPa/s and the reduction extent was 30.0%. As the methane concentration changed, $(dP/dt)_{max}$ was gradually decreased and its decrease extent was continuously increased. In particular, $(dP/dt)_{max}$ was the smallest and its reduction extent was greatest at 10% methane concentration (from 7.0 to 2.8 MP/s). Its drop percentage reached 52.0%. This fully demonstrates that the weakening effect of mist on mixture explosion reaction rates under different methane concentrations existed the obvious difference, and the weakening effect on the flame with lower explosion reaction rate was more obvious.

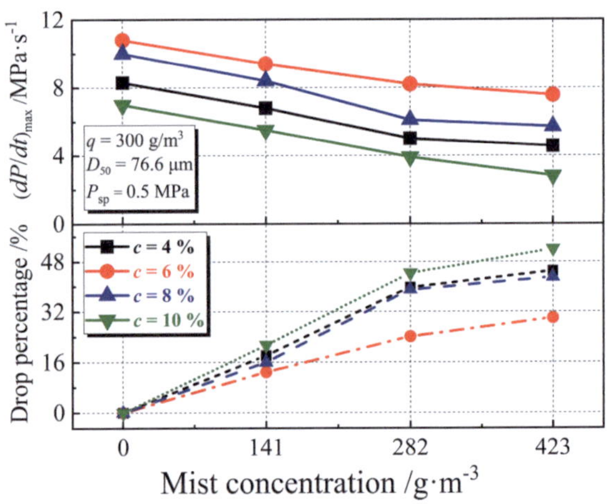

Fig. 4.11 $(dP/dt)_{max}$ and its drop percentage with the increases of methane and mist concentrations

4.2.5 Analysis of Flame Temperature

Figure 4.12 presents the comparison of temperature histories under four methane concentrations after adding mist ($Q = 282$ g/m^3). The temperature history exhibited the similar variation trend as the pressure history. T_{max} and its slope of temperature history all presented the trend of firstly rising and then reducing with the rise of methane concentration. As $c = 6\%$, the flame temperature was the highest ($T_{max} =$ 1382.0 °C) and its history rising rate was also the fastest. As the methane concentration changed, the temperature was gradually reduced and its history rise rate was also decreased. As $c = 10\%$, the flame temperature was the smallest ($\Delta T_{max} =$ 524.0 °C) and its history rising rate was the slowest. This demonstrates that the mist could reduce the flame temperature of mixture explosions and evidently affected by methane concentration. Figure 4.13 presents ΔT_{max} and its drop percentage with the methane and mist concentrations raised. The above parameters all experienced a variation trend of firstly reducing and then increasing as the methane concentration increased. Meanwhile, ΔT_{max} was gradually decreased and its drop percentage was ceaselessly improved as the mist concentration increased. As $c = 6\%$, ΔT_{max} was the highest and the decrease extent was the smallest. It was reduced from 1737.2 to 1105.0 °C and the reduction extent reached 36.4%. However, ΔT_{max} was the lowest and the decrease extent was the greatest at 10% methane concentration. It was reduced from 956.6 to 523.2 °C and the reduction extent reached 45.3%. This demonstrates that the increase in mist concentration could improve the inhibition effect of mist on mixture explosions, and the drop percentage in ΔT_{max} was more evidently for the weaker explosion flame.

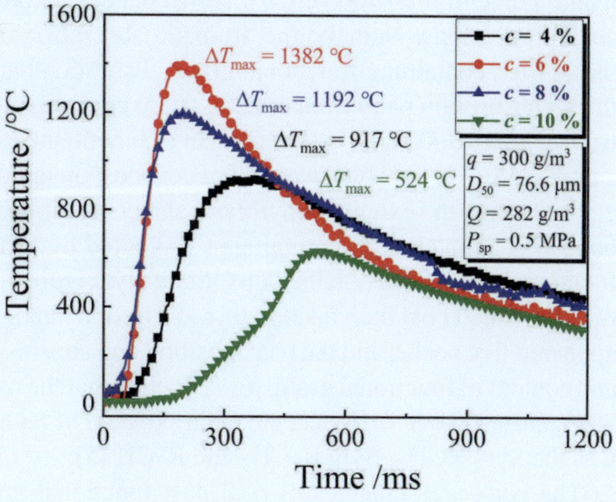

Fig. 4.12 Effect of methane concentration on the temperature histories under action of mist ($Q = 282$ g/m^3)

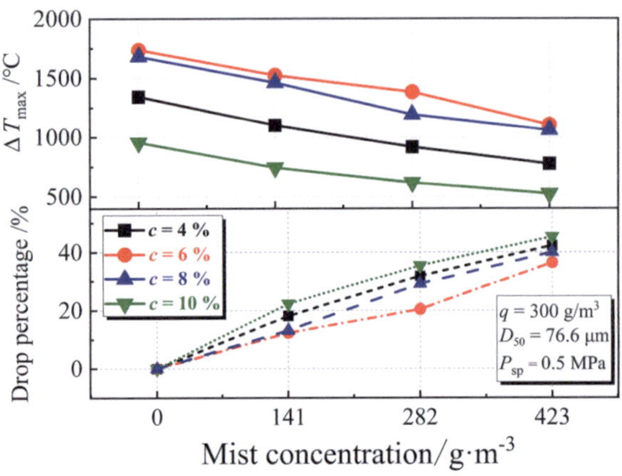

Fig. 4.13 ΔT_{max} and its drop percentage with the increases of methane and mist concentrations

4.2.6 Product Analysis

After the premixed gas was ignited, the flame sufficiently contacted with the micro scale coal dust suspended in the vessel and rapidly provided heat for its pyrolysis and combustion. Under the action of high temperature flame, volatile substance came from the pyrolysis of macromolecular functional group in coal dust was continuously precipitated [3]. After adding mist, the pyrolysis extent of functional group in coal dust could be obviously affected. An infrared spectroscopy analyzer was adopted to test the influence of mist concentration on the internal functional group of coal dust explosion product. Figure 4.14 shows the infrared spectrum within the wavelength range of 500–4000 cm^{-1} before coal dust explosion. It mainly included four absorption bands. Namely, the aromatic absorption band (range: 700–900 cm^{-1}), the oxygen containing functional group absorption band (range: 1000–1800 cm^{-1}), the fat absorption band (range: 2800–3000 cm^{-1}) and the hydroxyl absorption band (range: 3000–3600 cm^{-1}). The content of functional groups in the fatty structure was significantly affected by the extent of coal dust metamorphism and was a key parameter reflecting the extents of pyrolysis and combustion [5]. Hence, the fat absorption band was adopted and its total area was fitted by peak. And the indicator of each functional group was obtained quantitatively. Figure 4.15 presents the fitting history of unexploded coal dust fat structure absorption band. It clear that the fitting history appeared five peaks, and the peak position and area respectively represented the type and content of functional group [6, 7]. It clear that the main peak positions were 2865, 2850 and 2900 cm^{-1}. Hence, the main skeleton of fat macromolecule structure band was the Sym.RCH_3, Sym.R_2CH_2 and R_3CH [5].

The peak area changes of coal dust functional groups were obtained by peak fitting on the fatty structure bands of coal dust after explosion, and the extent to which fat structures in volatile matter participated in oxidation reactions was clarified. Figure 4.16 presents the fitting diagram of coal dust fat structure absorption band after explosion without mist. Compared with unexploded coal dust, the peak areas of

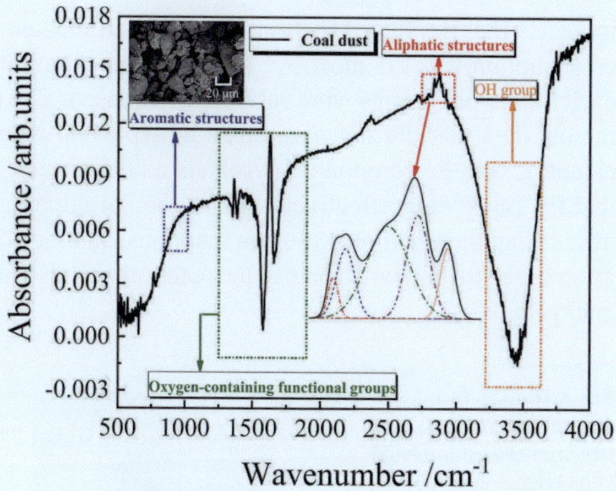

Fig. 4.14 The infrared spectrum within the wavelength range of 500–4000 cm^{-1} before coal dust explosion

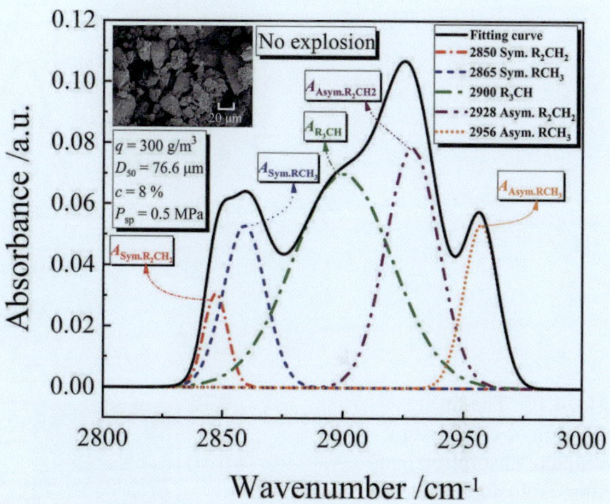

Fig. 4.15 The fitting history of unexploded coal dust fat structure absorption band

each functional groups in the fat structure all were decreased. Especially, the functional groups of R_3CH, Asym.R_2CH_2 and Asym.RCH_3 presented an evident decrease trend. This indicates that the functional groups in coal dust happened the pyrolysis oxidation reaction after explosion, and its content was obviously reduced [8]. The peak area in the fatty functional group of the explosion coal dust was evidently increased after adding mist, as can be seen in Fig. 4.17. It was because that the mist inhibited the pyrolysis oxidation reaction of volatile components in coal dust and its reaction rate was evidently reduced, leading to an increase in the content of functional group after coal dust explosion. Meanwhile, its peak areas under different mist concentrations after explosion was obtained by fitting the fat structure absorption band (in Fig. 4.18). The peak area of coal dust functional groups after explosion within the 2800–3000 cm^{-1} band all existed an evident change. As $Q = 0$ g/m^3, the peak areas of each functional group in the fat structure was decreased. This indicates that the fatty functional groups participated in explosion reaction and its consumption was the most. As the mist concentration increased, the peak areas of each functional groups were relatively increased. The contents of functional groups in coal dust that did not participate in explosion reaction were increased, and the extent of volatile component pyrolysis oxidation was decreased. This fully verifies that the mist could effectively inhibit the participation of functional group in coal dust during the mixture explosion reactions, and the contents of coal dust functional groups that did not participate in the reaction was also raised as the mist concentration enlarged.

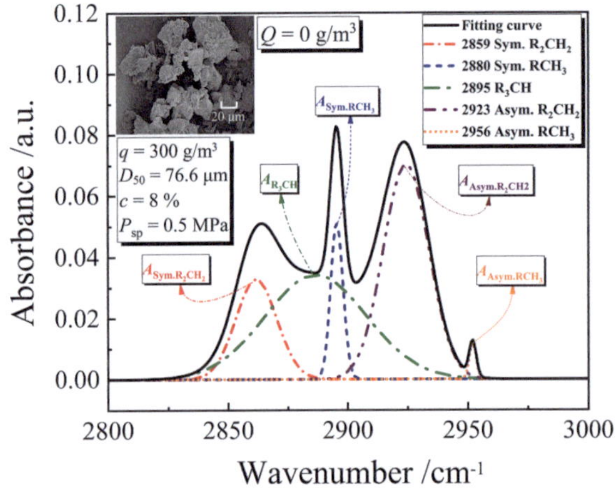

Fig. 4.16 The fitting diagram of coal dust fat structure absorption band after explosion without mist

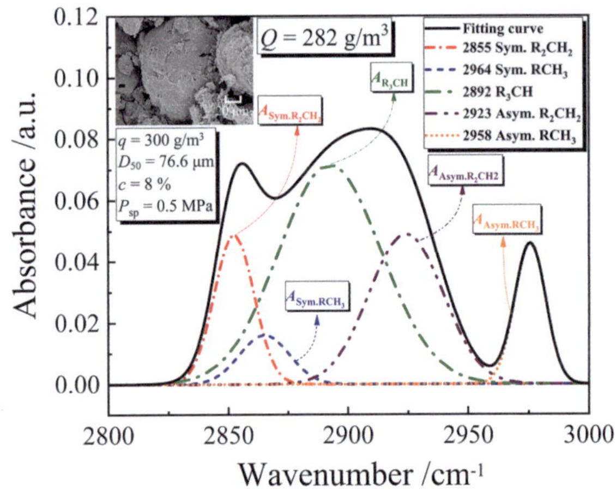

Fig. 4.17 The fitting diagram of coal dust fat structure absorption band after explosion under the action of mist ($Q = 282$ g/m^3)

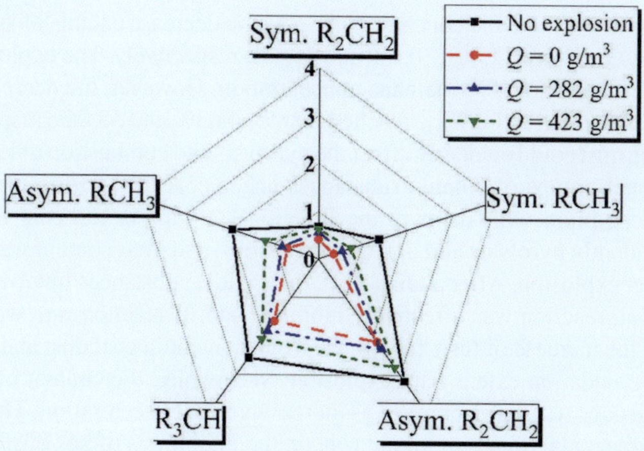

Fig. 4.18 The peak areas of coal dust functional groups under different mist concentrations after explosion

4.3 Conclusions

This research mainly aimed to the weakening degree and inhibiting mechanism of mist on methane/coal dust mixture explosion flames. By analyzing the reduction extents of explosion parameters after adding mist and the changes of explosion products in coal dust, as well as the correspondence between the explosion parameter inside the vessel, the weakening mechanism of mist on gas/dust explosions was revealed. The main conclusions are summarized as follows:

First, the pressure rise of gas/dust explosions presented an evident corresponding relationship as the flame propagated under the action of mist. However, the velocity history and (dP/dt) were evidently reduced compared with the case no mist. And the corresponding moment was also prolonged. The flames could be also obviously inhibited by mist and its propagation time inside the vessel was evidently increased. In particular, v_{max} was reduced and the corresponding moment was also prolonged. Due to the thermal absorption, the flame exhibited a dark-irregular structure. And the front exhibited the obviously thermal diffusion instability. The intensity of mixture explosions could be affected by the methane concentration, thereby affecting the weakening extent of mist on flame. Under the inhibitory effect of mist, the velocity history showed a variation regular of firstly rising and then reducing as the methane concentration increased.

Second, the mist could evidently weaken the intensities of methane/coal dust mixture explosions and its reaction rate. It was mainly reflected by the decreases of ΔP_{max}, (dP/dt) and ΔT_{max}. Meanwhile, the weakening effect could be ceaselessly enhanced after the mist concentration was increased. Besides, the weakening extent of mist on the mixture explosions was related to its intensity, and the inhibition effect on the weaker explosion flames was more significant. The explosion intensity was the strongest at 6% methane concentration. And the decrease extents of ΔP_{max}, $(dP/dt)_{max}$ and ΔT_{max} were 14.3%, 30.0% and 36.4%, respectively. The explosion intensity was the smallest at 10% methane concentration. However, the decrease extents of ΔP_{max}, $(dP/dt)_{max}$ and ΔT_{max} reached 22.8%, 44.3% and 52.0%, respectively.

Third, the mist could evidently affect the pyrolysis and combustion of large molecular functional groups of volatile substances inside coal dust during its explosion. Under the condition without mist, the functional groups in the coal dust underwent the evidently pyrolysis and oxidation reactions, and their content was evidently reduced after explosion. After adding mist, the volatile substances involved in pyrolysis oxidation reaction was effectively inhibited and its reaction rate was reduced, resulting in the increase of fatty functional group content in coal dust and a decrease of pyrolysis oxidation extent after explosion. Meanwhile, the content of unreacted functional groups was also improved as increasing mist concentration. The chemical effect also given play to an important role on the inhibitions of gas/dust explosions except for the physical effect.

For industrial explosion protection, some application suggestions were proposed according to the current research conclusions: (1) To effectively reduce the pyrolysis and oxidation reactions of macromolecular functional groups in the coal dust, the chemical inhibitor that could effectively inhibit the pyrolysis of volatile matter in coal dust should be dissolved into the mist. (2) Flammable gas contained in dust should be kept as far away as possible from the concentration corresponding to the maximum intensity of mixture explosions. (3) The explosion inhibition equipment that could quickly generate high water mist flux and small water mist diameter should be developed.

References

1. Cao XY, Ren JJ, Bi MS, Zhou YH, Li YM. Experimental research on the characteristics of methane/air explosion affected by ultrafine water mist. J Hazard Mater 2017; 324: 489–497.
2. Cao XY, Ren JJ, Zhou YH, Wang QJ, Gao XL, Bi MS. Suppression of methane/air explosion by ultrafine water mist containing sodium chloride additive. J Hazard Mater 2015; 285: 311–318.
3. Cao XY, Wei HY, Wang ZR, Fan LT, Zhou YQ, Wang Z. Experimental research on the inhibition of methane/coal dust hybrid explosions by the ultrafine water mist. Fuel 2023; 331: 125937.
4. Cao XY, Wang ZR, Lu YW, Wang Y. Numerical simulation of methane explosion suppression by ultrafine water mist in a confined space. Tunn Undergr Sp Tech 2021; 109: 103777.
5. Li QZ, Yuan CC, Tao QL, Zheng YN, Zhao Y. Experimental analysis on post explosion residues for evaluating coal dust explosion severity and flame propagation behaviors. Fuel, 2018; 215: 417–428.
6. Nie BS, Yang LL, Ge BQ, Wang JW, Li XC. Chemical kinetic characteristics of methane/air mixture explosion and its affecting factors. J Loss Prev Process Ind 2017; 49: 675–682.
7. Nie BS, Yang LL, Wang JW. Experiments and Mechanisms of gas explosion suppression with foam ceramics. Combust Sci Technol 2016; 188(11–12): 2117–2127.
8. Song L, Liu Z, Zhao E, Qian J, Ali M. A study on the FTIR spectra of pre-and post-explosion coal dust to evaluate the effect of functional groups on dust explosion. Process. Saf Environ Prot. 2019; 130: 48–56.

Chapter 5
Research on the Venting Characteristics of Methane and Coal Dust Two-Phase System Mixture Explosions

5.1 Experimental Apparatus and Procedures

Figure 5.1 presents a structure diagram of experiment apparatus for studying the venting characteristics of methane and coal dust mixture explosions. It mainly consists of the explosion venting vessel, an ignition system, a gas distribution system, a dust spraying system, a pressure acquisition system, a temperature acquisition system, an image acquisition system, a program control and data acquisition system, as well as the corresponding accessory.

The explosion venting vessel ($H \times D$: 400 mm × 318 mm) was a cylindrical structure with an venting port ($L \times H$: 180 mm × 100 mm). And its window was made of two high borosilicate tempered glasses ($\delta \times D$: 30 mm × 100 mm). The main body of vessel was made of stainless steel, and its upper and lower ends were sealed with flanges with 30 mm thickness. The gas/dust mixture substances were ignited by a high voltage discharge method (Ignition energy (E) = 8 kV). The ignition electrode was located in the center of vessel and the gap between the two electrode wires was 6 mm. The premixed gas was prepared in the premix tank and the dust injection pressure was controlled by changing its initial pressure. Hartmann dust spraying device was used to realize the coal dust raising process. And it was installed at the bottom end of vessel and connected to the premix tank through a solenoid and check valves. The airflow formed a vortex from bottom to top as it passed through the nozzle cover due to the blocking action, lifting the coal dust in the groove and forming a coal dust cloud. Four high-frequency pressure transmitters (the range: 0–2.5 MPa; the acquisition accuracy: 0.25%) were placed at the center of vessel upper flange, the wall surface of venting pipe, and 500 and 1000 mm horizontal positions distance from the venting port (L) to collect the changes of pressure and temperature with time. Besides, a high-speed camera (Phantom V1212) was adopted to record the propagation characteristics of venting flame under different working conditions (4000 fps). A high-speed infrared thermal imager (TELOPS FAST-IR) was adopted to record the temperature field distribution change of venting flame (1000 fps). The

Z. Wang and X. Cao, *Mechanism of Composite Explosion Inhibition and Venting Protections*, https://doi.org/10.1007/978-981-95-4809-5_5

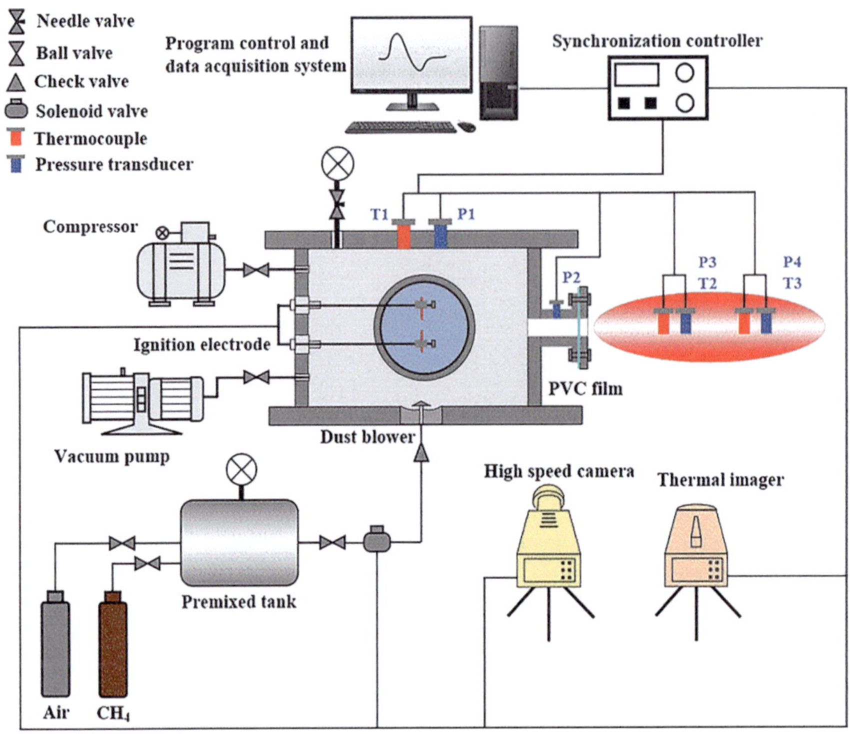

Fig. 5.1 Structure diagram of experiment apparatus

order of dust spraying, ignition and explosion parameter acquisition was achieved through a program control and data acquisition system.

The mixture gases with three kinds of methane concentrations ($c = 4$, 6 and 8%) were prepared inside the premix tank. The parameters of coal dust were the important influencing factors for the venting characteristics of gas/dust mixture explosions. Through the explosion venting experiment under different coal dust parameters, the explosion venting parameter was the greatest as the dust concentration was 300 g/m^3. It indicates that the coal dust concentration reached the optimal value inside the vessel. Meanwhile, the explosion intensity presented an increasing trend as the coal dust diameter decreased. To ensure better suspension and distribution of coal dust inside the vessel, the coal dust with determined diameter ($D_{50} = 35.59$ μm) was placed inside the dust groove to ensure the dust concentration inside the vessel after dust raising ($q = 300$ g/m^3). The coal block was crushed using a crusher and refined by a ball mill. And its component content was measured by the industrial analysis ($M_{ad} = 8.5\%$; $A_{ad} = 6.3$; $V_{ad} = 42.8\%$; $FC_{ad} = 42.0\%$). The morphological characteristics of coal dust was analyzed by scanning electron microscope (×2000), as shown in Fig. 5.2. And the diameter distribution of coal dust was measured using the laser particle size analyzer (Mastersizer 2000). It clear that the morphology and

Fig. 5.2 Columnar distribution of coal dust particle size and its microstructure

structure of coal dust presented a uniform and smooth states, and its diameter history approximated a normal distribution.

Besides, its thermogravimetric history under the high temperature was measured, as shown in Fig. 5.3. At a heating rate of 10 K/min, the coal dust mainly underwent three stages: the heat absorption, weight loss and stability. The pyrolysis occurred at 295 °C. And the TG history began to decrease and the weight loss rate was gradually increased (the acceleration stage). At 578 °C, the weight loss rate was the fastest and its corresponding DTG history reached the peak value. After that, the weight loss rate was decreased (the deceleration stage). As the temperature increased to 710 °C, the weight loss of coal dust reached a stable state. The positive and negative values represented respectively the heat release and absorption in the DSC history. During the entire thermal analysis, the DSC history presented a continuous endothermic process. At 568 °C, the DSC history reached the peak value of heat absorption. It concludes that a violent oxidation reaction occurred after gases volatilized from pyrolysis as the temperature reached 568 °C [1].

Before the experiment, the determined vacuum degree inside the vessel was achieved by a vacuum pump. The dust injection pressure of premix tank was determined by the calculation. The premixed gas with a determined concentration was realized after dust spraying and the initial pressure was 0.10 MPa as the mixture was ignited. According to the experiment, the intensity of methane and coal dust mixture explosions was the greatest ($P_{max} = 0.64$ MPa) at 0.40 MPa dust injection pressure. Hence, this injection pressure was adopted during the experiment. The time of dust injection was controlled by a procedure ($t = 700$ ms) [2]. Subsequently, the mixture was ignited (ignition duration: 100 ms). During the experiment, the strength of film rupture was 0.32 MPa. The explosion flame inside the vessel continued to development and spread after igniting. The flame was released to the exterior of vessel after the pressure exceeded the strength of film, causing the sharp increases of pressure and temperature. To ensure the accuracy and reliability of the experiment, each group of experiment was repeated 3–4 times under the same working conditions. Through

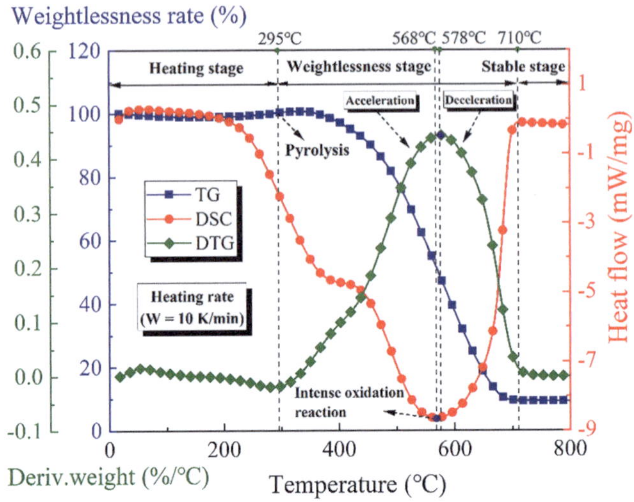

Fig. 5.3 Thermogravimetric history of coal dust

the repeated experiments, the maximum pressure deviation values at four points was less than 5.6%.

5.2 Results and Discussions

5.2.1 Corresponding Relationship of Explosion Venting Parameters

Figure 5.4 shows the corresponding relationship between the pressure rise and flame propagation inside and outside the vessel during the explosion venting ($c = 6\%$). The mixture of methane and coal dust was ignited by the high-voltage discharge method. As the explosion reaction progressed, the rise moment of pressure inside the venting pipe ($t = 59.0$ ms) was prior to the pressure inside the vessel ($t = 62.5$ ms) due to the acceleration effect. And the rising rate of pressure history was also obviously increased compared with the interior. Especially P_{max} reached 0.43 MPa and it was increased by 34.88%. At 105.5 ms, the venting film ruptured as the internal pressure of venting pipe ($P = 0.33$ MPa) was slightly higher than the strength of venting film ($P = 0.32$ MPa). The flame was released to the vessel exterior and presented an irregular structure at the film rupture moment. And it began to diffuse and stretch along the radial direction. With the acceleration propagation of flame in the early stage, the surface area of flame front was continuously increased and its propagation velocity appeared the radial component under actions of high pressure difference and resistance. Thereafter, the venting flame was gradually formed the

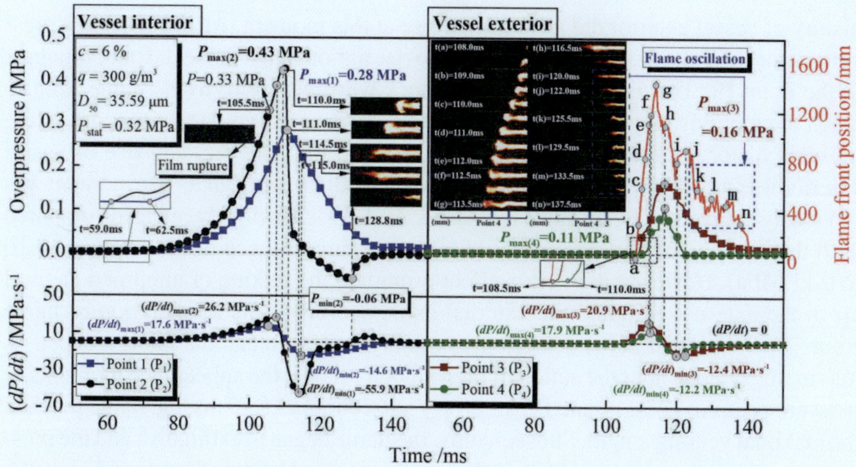

Fig. 5.4 The corresponding relationship between the flame propagation and pressure rise inside and outside the vessel during the explosion venting ($c = 6\%$)

mushroom cloud-like structure and large-scale vortex structures under the actions of diffusional-thermal instability, Kelvin–Helmholtz instability and flow eddies [3]. At 110.0 ms, the internal pressure of venting pipe reached the peak value. However, the peak pressure of vessel interior appeared slightly later (t = 111.0 ms). With the attenuation of venting intensity, the vortex motion of flame was decreased and disappeared eventually, the flame developed in a "spherical" structure and its surface area was continuously decreased.

Compared with the vessel interior, the pressure drop rate of venting pipe interior was more significant and the $(dP/dt)_{min(1)}$ reached -55.9 MPa/s^{-1}. Meanwhile, the flame near the outer layer continued to disappear with the enhancement of heat dissipation. After that, it presented a "strip" structure and tended to upward due to the buoyancy. The flame began to extinguish after the venting flame propagated to the maximum length distance from the venting port. The pressure at the venting port was decreased sharply with the extinction of venting flame. And its history appeared a larger negative value ($P_{min} = -0.06$ MPa) and then restored to normal pressure in the later stage. This is because that the venting rate continued to increase and ultimately exceeded the rise rate of pressure generated from the gas explosion as the flame propagated in the later stage, further resulting in the rapid decrease of (dP/dt) history [4]. Meanwhile, the flame continued to accelerate propagate forward due to the action of inertia as the pressure inside the vessel was closed to the environmental pressure, causing the further decrease of pressure near the venting port [5]. Therefore, the pressure history of venting pipe appeared a larger negative value compare with the vessel interior.

The propagation of venting flame and the rise of pressure outside the vessel also presented an obvious corresponding relationship. After the venting film ruptured, the flame horizontally propagated along to the venting port. However, the pressure history of vessel exterior did not begin to rise at this moment. At 108.5 ms, the pressure history at a distance of 500 mm began to rise although the flame did not propagate to the Point P_3. This is because that the shock wave was prior to the flame front to reach the monitoring point during the venting. As the venting flame propagated, the pressure history at Point P_3 began to rise at 110.0 ms. The pressure rise of vessel exterior was more dramatic compared with the vessel interior, but its rise moment was obviously delayed as the distance increased. And it was obviously reduced compared with the interior and further decreased as the venting distance increased (from 0.16 to 0.11 MPa). However, the vortex effect promoted the mixing of unburned gas and air in the external field and the external explosion drove the acceleration of flame front. The horizontal flame velocity was rapidly increased and the flame reached its maximum value. Under the action of high velocity flame, the appearance moments of pressure peaks at Point P_3 and P_4 were very adjacent. At 113.5 ms, the flame reached the maximal venting length. Subsequently, the flame began to extinguish and the pressure appeared the decrease. Due to the continuous reduction of venting flame, the pressure at Point P_4 was obviously decreased and reached to normal pressure earlier compared with Point P_3. During the extinguishing stage (t = 120.0–137.5 ms), the flame front appeared an obvious oscillation phenomenon. Namely, the increase and decrease changes of venting flame length. It was because that the reflection and superposition of pressure wave caused by vessel front and rear wall surfaces, leading to a flame-pressure wave coupling effect between the reverse pressure wave and flame at venting port, and further resulting in the high-frequency oscillation of flame along the horizontal direction. The interaction between the flame and pressure wave promoted the flame transition from the laminar to turbulent flow. As the oscillation developed, the turbulence intensity continued to increase. After that, the flame was gradually decreased in a "strip" structure and disappeared accompanied by the pressure reaching normal pressure.

5.2.2 Propagation Characteristics of Venting Flame

Figure 5.5 presents the comparison of venting flame propagation process under different methane concentrations ($q = 300$ g/m^3, $D_{50} = 35.59$ μm, $P_{stat} = 0.32$ MPa). The moment in the figure was the relative to the moment of venting film rupture. At 0.5 ms, the venting flame propagated horizontally after the film ruptured and did not presented an obvious difference in the morphology and structure. As the venting flame propagated, the venting flame at 6% methane concentration was slightly longer and wider. Due to the resistance effect, a clear "mushroom" structure appeared on the flame front at 6 and 8% methane concentrations. And the "mushroom" structure at 6% methane concentration was more significant. However, the flame only propagated in a "strip" structure at 4% methane concentration. As the venting continued to propagate, the flame showed an obvious "spherical" structure at 6 and 8% methane concentrations.

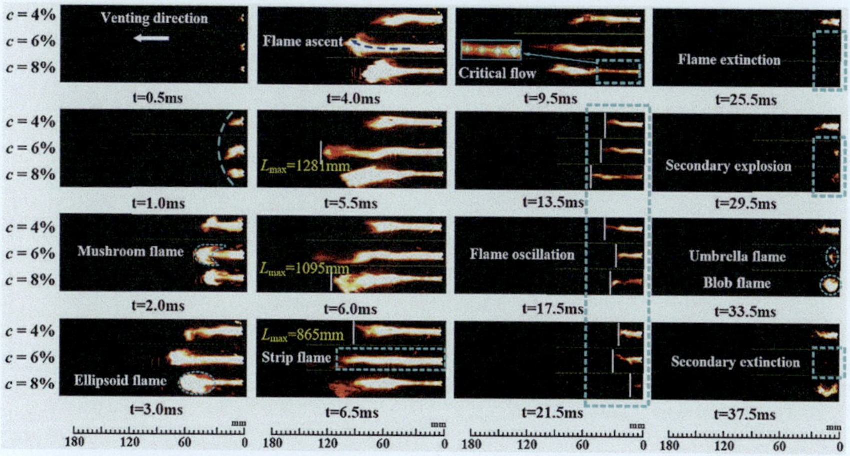

Fig. 5.5 Comparison of the venting flame propagation process

Combined the Figs. 5.6 and 5.7, the venting flame was the fastest at 6% methane concentration, indicating the corresponding maximal explosion venting intensity. The explosion venting intensity corresponding to 4% methane concentration was the smallest. In the later stage, the flame propagated in a "strip" structure and tended to float upward due to buoyancy. However, the flame exhibited a longer venting distance at 6% methane concentration, followed by 8% methane concentration. The venting distance was the shortest at 4% methane concentration, but its discharge duration was the longest during the extinguishing stage. The flame oscillation was more significant at 6 and 8% methane concentrations. This indicates that the oscillation extent was related to the gas explosion intensity inside the vessel. Besides, the venting flame appeared an obvious secondary venting phenomenon at 6 and 8% methane concentrations. Namely, a "spherical" flame reappeared and vented to the vessel exterior after the flame disappeared. This is because that the negative pressure effect of venting port promoted the reflux of high temperature gas and fresh air. It was ignited by the high temperature gas after the mixture concentration inside the vessel reached the explosion limit, further resulting in a secondary venting phenomenon [6]. And the phenomenon was more significantly and the venting flame area was larger at 8% methane concentration. The appearance moment of secondary venting flame under two methane concentrations were 32.3 ms and 35.8 ms, respectively. And the corresponding maximum venting length reached 173 mm at 40.3 ms and 269 mm at 43.3 ms. This indicates that the secondary venting flame had the significant high temperature damage to the local area of venting port. Besides, the secondary venting flame presented respectively the "umbrella" and "blob" structures at 6% and 8% methane concentrations.

Fig. 5.6 Comparison of the venting flame front position

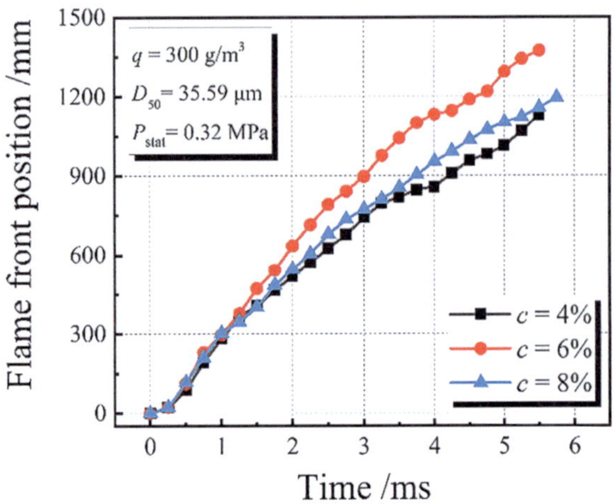

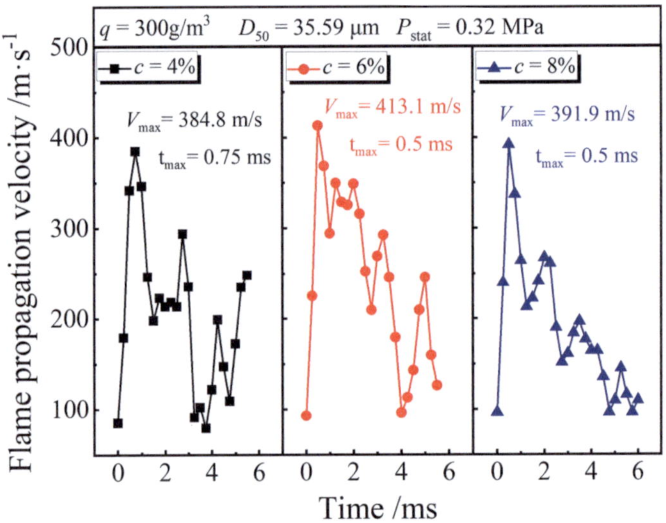

Fig. 5.7 Comparison of the venting flame propagation velocity

5.2.3 Analysis of Explosion Venting Pressure

Figure 5.8 shows the comparison of pressure histories at four positions during the venting of mixture explosions under the determined the methane and coal dust parameters (c = 6%, P_{stat} = 0.32 MPa). It clear that the pressure history of four positions showed a change trend of increasing firstly and then decreasing as the venting progressed. And the maximal pressure appeared at the venting port, followed by the vessel interior. As the distance increased, the venting pressure was gradually decreased. Compared with the venting port, P_{max} was decreased by 62.79% and 74.42%, respectively. It was because that the flame was accelerated and the shock wave intensity was evidently increased under the action of the venting pipe. After

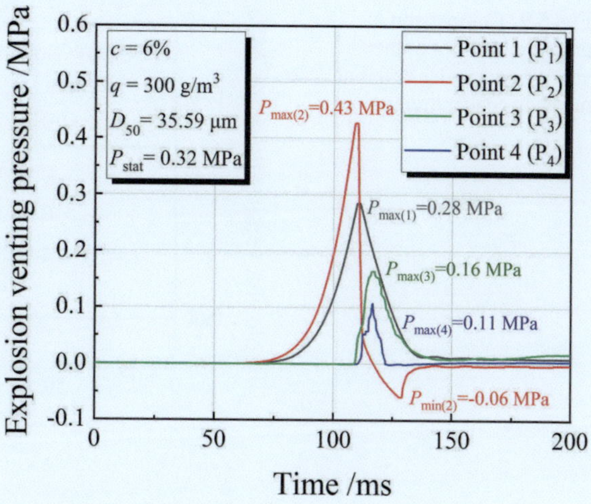

Fig. 5.8 Comparison of pressure histories at four positions during mixture explosion venting ($c = 6\%$, $P_{stat} = 0.32$ MPa)

the film ruptured, the flame was released to the open space outside the vessel under the actions of high-pressure and high-speed airflow. The expansion and reflected compression waves were formed at the edges on both sides of venting port, and the both them intersected and converged alternately. Under the actions of turbulence instability and friction mixture, the intensity of shock wave was evidently attenuated, and the flame near venting port presented an evident under-expansion periodic flow structure, as shown in Fig. 5.5. Meanwhile, the areas of shock wave and venting flame continued to increase and generated the radial vortex motion due to the lack of wall constraint. Under the effects of air resistance and heat dissipation, the shock wave was attenuated and the pressure was decreased as the venting distance increased. Besides, the rise moment of vessel interior pressure history was evidently delayed although its rise rate was faster compared with the vessel interior. And the rise moment of pressure history near the venting port was slightly advanced and appeared a larger negative value.

Figure 5.9 presents the comparison of pressure histories at the venting port under different methane concentrations ($q = 300$ g/m^3, $D_{50} = 35.59$ μm, $P_{stat} = 0.32$ MPa). As the methane concentration increased, the venting pressure was first increased and then decreased. As $c = 6\%$, the venting pressure was the largest ($P_{max} =$ 0.43 MPa). And the slope of pressure history was also the greatest, followed by 8% methane concentration. Its slope was the smallest at 4% methane concentration. Meanwhile, the negative value also presented an obvious corresponding relationship of increasing firstly and then decreasing. Combined with Table 5.1, the venting pressure all presented the same change trend under different methane concentrations, and the maximal venting pressure all appeared at the venting port. The pressure at the place away from the venting port was sharply decreased. Besides, the negative value of pressure history at the venting port was related to the explosion venting intensity and was increased with its enhancement.

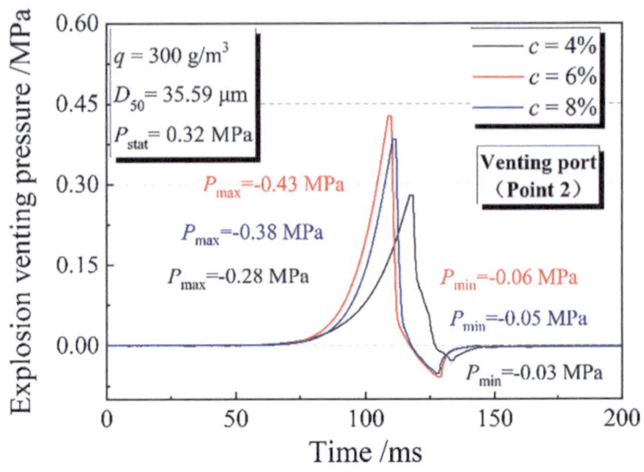

Fig. 5.9 Comparison of pressure histories at the venting port under different methane concentrations

Table 5.1 Effect of methane concentration on the explosion venting pressure and their appearance moment

Monitoring position	Experimental data	Methane concentration		
		4%	6%	8%
Point 1	P_{max}/MPa	0.21	0.28	0.27
	Appearance moment/ms	118.75	110.52	112.08
	Rise moment/ms	67.11	63.52	66.03
Point 2	P_{max}/MPa	0.28	0.43	0.38
	Appearance moment/ms	117.53	109.75	111.25
	P_{min}/MPa	−0.03	−0.06	−0.05
	Appearance moment/ms	133.23	128.26	127.61
	Rise moment/ms	67.75	63.75	66.52
Point 3	P_{max}/MPa	0.09	0.16	0.12
	Appearance moment/ms	125.51	116.75	118.75
	Rise moment/ms	117.25	108.05	111.75
Point 4	P_{max}/MPa	0.02	0.11	0.06
	Appearance moment/ms	125.25	117.05	122.75
	Rise moment/ms	117.75	109.25	112.35

5.2.4 Analysis of Venting Pressure Rising Rate

The intensity of explosion venting could be directly reflected in the rise rate of venting pressure. Figure 5.10 presents the comparison of pressure rising rate histories at four positions during the venting under the determined the methane and coal dust parameters ($c = 6\%$). The (dP/dt) history presented the same change trend as the pressure history with the increase of venting distance. Namely, it showed a change trend of increasing firstly and then decreasing. And the change was the most

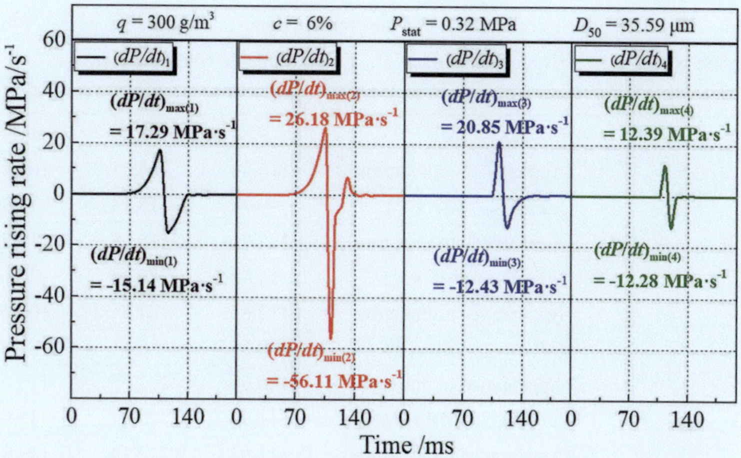

Fig. 5.10 Comparison of venting pressure histories under different methane concentrations

significantly at the venting port. The $(dP/dt)_{max}$ and $(dP/dt)_{min}$ reached 26.2 MPa/ s^{-1} and -56.1 MPa/s^{-1} respectively. Table 5.2 shows the $(dP/dt)_{max}$ and $(dP/dt)_{min}$ and their appearance moments at the four points under different methane and coal dust parameters. It clear that the explosion venting parameters all presented the same change regular with the increase of venting distance. This indicates that the reaction rate of mixture explosions at the venting port was the most intense and the venting flame propagation was the fastest, and the pressure change was also most significantly. Figure 5.11 shows the comparison of pressure rising rate histories at the venting port under different methane concentrations ($q = 300$ g/m^3, $D_{50} = 35.59$ μm, $P_{stat} = 0.32$ MPa). It clear that the (dP/dt) history also presented a change trend of increasing firstly and then decreasing with the increase of methane concentration. The pressure rise rate at 6% methane concentration was largest ($(dP/dt)_{max} = 20.9$ MPa/ s^{-1}). The moment for pressure history began to rise and the appearance moment of $(dP/dt)_{max}$ were obviously advanced. Besides, the negative value corresponding to (dP/dt) history was also the smallest. Combined with Table 5.2, the rise rate of venting pressure at the four points all presented the same change regular with the increase of methane concentration.

5.2.5 *Analysis of Venting Flame Temperature*

The distribution and variation of the venting flame temperature field could intuitively reflect the venting characteristics and disaster-mechanism of mixture explosions. Figure 5.12 shows the changes in flame propagation image and temperature field distribution during the explosion venting at the atmospheric pressure ($c = 6\%$, $P_{stat} =$ 0.32 MPa). It clear that a good correspondence was presented between the two. After the film ruptured, a high temperature flame appeared at the local area of venting port at

Table 5.2 Effect of methane concentration on the pressure rising rate and their appearance moment

Monitoring position	Experimental data	Methane concentration		
		4%	6%	8%
Point 1	$(dP/dt)_{max}/(\text{MP s}^{-1})$	9.82	17.29	16.91
	Appearance moment/ms	113.54	105.51	107.12
	$(dP/dt)_{min}/(\text{MP s}^{-1})$	−9.77	−15.14	−14.35
	Appearance moment/ms	125.71	115.58	117.52
Point 2	$(dP/dt)_{max}/(\text{MP s}^{-1})$	14.48	26.18	24.10
	Appearance moment/ms	112.81	104.55	106.25
	$(dP/dt)_{min}/(\text{MP s}^{-1})$	−29.54	−56.11	−51.07
	Appearance moment/ms	120.91	111.73	113.67
Point 3	$(dP/dt)_{max}/(\text{MP s}^{-1})$	11.91	20.94	15.49
	Appearance moment/ms	121.92	112.13	114.91
	$(dP/dt)_{min}/(\text{MP s}^{-1})$	−6.98	−12.41	−10.17
	Appearance moment/ms	133.14	121.89	125.52
Point 4	$(dP/dt)_{max}/(\text{MP s}^{-1})$	2.94	12.39	6.67
	Appearance moment/ms	121.01	112.23	117.11
	$(dP/dt)_{min}/(\text{MP s}^{-1})$	−2.15	−12.28	−4.76
	Appearance moment/ms	135.53	120.12	132.78

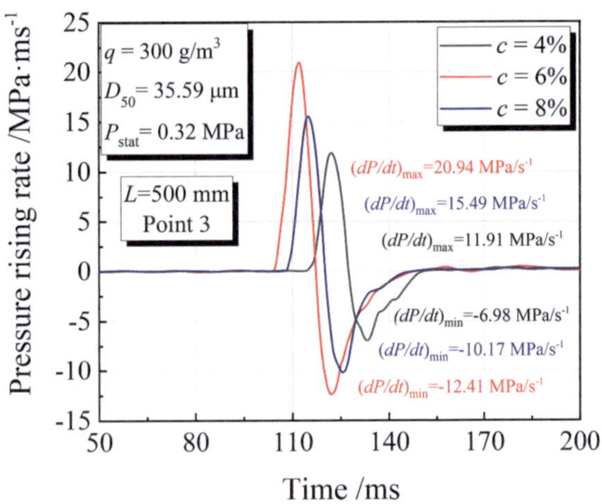

Fig. 5.11 Comparison of (dP/dt) histories at four positions during the venting ($c = 6\%$)

1 ms. The venting flame was accelerated under the high pressure difference ($t = 2.0$–5.0 ms). At this moment, the venting flame image was coincided with the temperature field distribution contour. Under the actions of resistance and drag force, the flame showed an upward process and the temperature field presented the same change trend at 4.0 ms. At 5.0 ms, the temperature of flow field reached the peak. As the venting

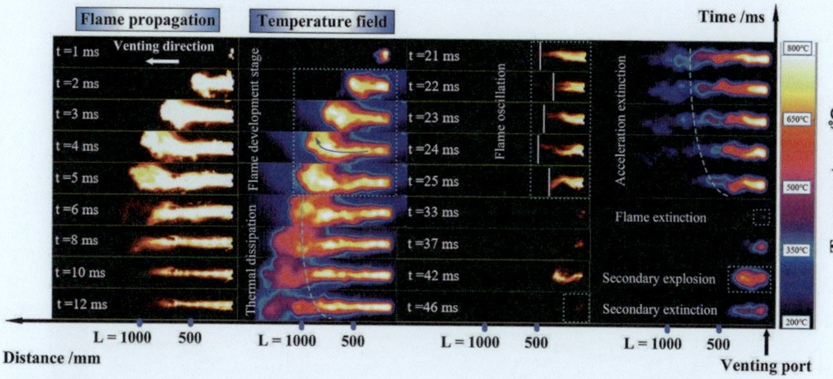

Fig. 5.12 Changes of flame propagation image and temperature field distribution during the venting at the atmospheric pressure ($c = 6\%$, $P_{stat} = 0.32$ MPa)

progressed, the length of venting flame continued to decrease due to the reduction in the venting pressure of vessel interior. During the flame extinguishing stage ($t =$ 5–25 ms), the temperature field distribution area was evidently larger than venting flame image and became discontinuous in the later stage, and the difference in width was also clear. Especially the phenomenon was more significant during the secondary venting. It was because that the area of venting flame was continuously decreased due to the heat dissipation. However, the extinguished flame still had a higher temperature. In the later stage, the flame appeared an obvious oscillation and the corresponding temperature field also presented the same change ($t = 21.0$–25.0 ms). Afterwards, the temperature field at the venting port disappeared with the extinguishment of venting flame. However, an obvious secondary venting phenomenon occurred at 33.0 ms. And the flow field at the venting port appeared a local high temperature again and the its distribution range was 500–650 °C. Compared with the venting flame, the distribution region of temperature field was larger. And the flame eventually extinguished at 46 ms.

Figure 5.13 shows the corresponding relationship between the temperature field and its history along the horizontal direction of venting port under three different methane concentrations. It can be seen that the T_{front} and T_{max} of three methane concentrations all presented a change trend of increasing firstly and then decreasing. And the T_{max} all appeared in the central area of spherical flame. Among them, the temperature of venting flame was the highest and the appearance moment was earliest under 6% methane concentration. As $t = 3$ ms, the T_{max} reached 979 °C. And the distribution area of temperature field was also the largest. Under 4% methane concentration, the T_{max} of venting flame was smallest and appearance moment was the latest. As $t = 5$ ms, the T_{max} reached 910 °C. For the 8% methane concentration, T_{max} was 940 °C. Compared with the 4 and 8%, the venting intensity of methane and coal dust mixture explosions was the most intense under 6% methane concentration, resulting in the higher flame temperature. It indicates that the temperature distribution of venting flame flow field was related to the explosion intensity inside the vessel.

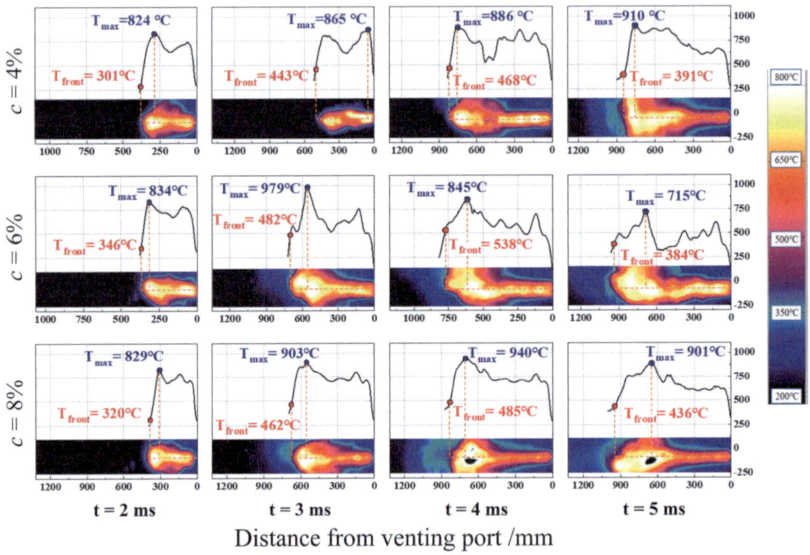

Fig. 5.13 The distribution and change of temperature field under three different methane concentrations

5.2.6 Mechanism Analysis of Explosion Venting

The coal dust entered the venting vessel interior under the action of pressure difference and the gas/dust two-phase system was uniformly dispersed. Since the ignition temperature of methane was far smaller than that of coal dust, the continuous premixed gas was ignited under the action of high-voltage discharge before the coal dust combustion and the flame continuously spread and propagated. During the propagation, the heat was ceaselessly transferred to coal dust. The flame contacted with the coal dust uniformly dispersed inside the vessel and provided the heat for its pyrolysis and combustion in the forms of convection and thermal radiation. Under the action of high temperature flame, the temperature of coal dust particles was increased and the heat was rapidly transferred from the surface to the interior [7]. This resulted in that the volatile combustible substances in coal dust continuous occurred the pyrolysis reaction, and substances with large molecular chemical bond structures broken into a large number of small molecular gaseous (N_2, O_2, CO_2, CO, CH, etc.) and liquid products (Hydrocarbons) [8]. The precipitated volatile matter expanded rapidly inside the coal dust, resulting in an obvious increase of coal dust volume. Volatile gases were released from the surface of coal dust with the increase of interior pressure, further forming the pore structure after coal dust explosion [9]. The combustion of volatile gases on the surface of coal dust could further promote the explosion reaction of methane and provide the heat for the combustion reaction of coal dust, which resulted in that the more coal dust particles occurred the pyrolysis reaction and released more volatile substances [10]. After the volatile matter was separated out, the remaining coke would undergo the multiphase reactions with O_2,

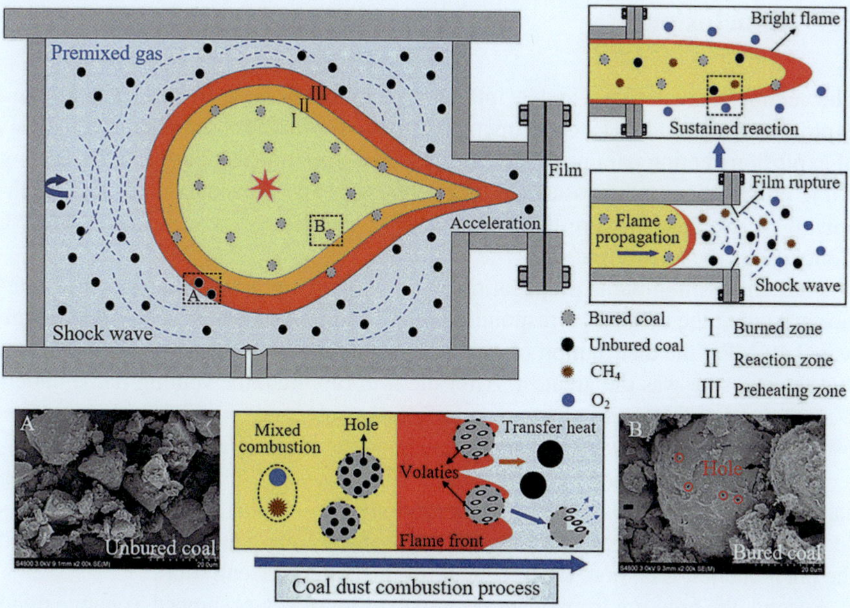

Fig. 5.14 Mechanism analysis of gas/dust mixture explosion venting

CO and H_2O under the action of high temperature flame ($C + O_2 \rightarrow CO_2$, $2C + O_2 \rightarrow 2CO$, $C + CO_2 \rightarrow 2CO$, $C + H_2O \rightarrow CO + H_2$) [11] (Fig. 5.14).

The gas and dust two phase systems resulted in a higher explosion reaction rate inside the vessel. After the methane and coal dust were ignited in the center of the vessel, the explosion flame continuously spread and propagated toward the wall surface of vessel, accompanied by a rapid heat and mass transfer processes between gas and dust two phases. Under the action of venting pipe, the shock wave and flame was accelerated toward the venting port and propagated along the pipeline direction. Affected by the shock wave, the film would rupture as the venting pressure exceeded its strength. After that, the shock wave and venting flame continuously propagated horizontally along the direction of venting pipe. During the venting, the unburned premixed gas and coal dust also would continue to occur the explosion reaction accompanied by the heat and mass transfers. Macroscopically manifested as the bright venting flame. Meanwhile, the reverse propagation of shock wave would further enhance and affect the venting of flame due to the obstruction of vessel wall surface [12]. After the flame was released, the pressure inside the vessel continued to increase due to the rise of temperature, leading to a small amount of unburned substance being released again and contacted with oxygen in the environment, which resulted in a secondary venting phenomenon at the venting port under the action of high-temperature gases.

5.3 Conclusions

The venting characteristics and its influence factor of methane and coal dust mixture explosions were studied by experiments. Combined with the comprehensive analysis of explosion venting parameter, the corresponding relationship between the venting flame propagation and pressure rise was obtained. The influence factor and mechanisms of gas/dust mixture explosion venting characteristics were also revealed. Conclusions are as follows:

First, the venting flame propagation and pressure rise of vessel interior and exterior presented the evident corresponding relationship. As the dust spraying pressure was 0.40 MPa, the distribution of coal dust inside the vessel was the most uniform and the pressure was the greatest. After the film ruptured, the venting flame underwent the evolution process of "mushroom", "spherical" and "strip" structures. As $c =$ 6%, the venting flame was the longest and its structure change was the most significant, and its propagation velocity was also the fastest. During the extinguishing stage, the flame appeared an obvious oscillation phenomenon. Especially an obvious secondary venting phenomenon appeared at 6 and 8% methane concentrations. And the phenomenon was more significant at 8% methane concentration.

Second, the pressure from the interior to exterior of vessel showed a change trend of increasing firstly and then decreasing as the venting progressed. The pressure at the venting pipe was the highest and its change was also most significant, even appeared a larger negative value. Followed by the vessel interior. As the explosion intensity increased, the P_{min} was also showed a decreasing trend. The corresponding (dP/dt) history also presented the same change trend. The venting pressure was the greatest at 6% methane concentration, followed by 8% methane concentration, and the venting pressure was the smallest at 4% methane concentration.

Third, the venting flame image presented a good correspondence with the temperature field distribution. Under the action of high pressure difference, the flame was acceleratingly released after the film ruptured. The highest temperature appeared in the central area of spherical flame. Due to the heat dissipation, the temperature at the front end of venting flame continued to decrease. And the extinguished flame still had a higher temperature. At 6% methane concentration, the flame temperature was the highest and the appearance moment of T_{max} was the earliest, followed by 8% methane concentration. The secondary venting resulted in the recurrence of high temperature flame at the venting port and the temperature reached 500–650 °C, and was affected by the methane concentration inside the vessel.

To better realize the effective protection gas/dust mixture explosions, three suggestions are proposed based on present experiments: (1) The concentration of flammable gas in gas/solid mixture substances should be reduced as much as possible to avoid secondary venting and reduce high temperature damage. (2) The combination method of venting and spraying inhibitor should be adopted to protect against gas/dust mixture explosions, reducing the damage caused by the high temperature flame. (3) Equipment and construction should be kept away from venting port to avoid damage caused by shock wave.

References

1. Zhang XY, Yu JL, Gao W, Zhang DW, Sun JH, Guo S, Dobashi R. Effects of particle size distributions on PMMA dust flame propagation behaviors. Powder Technol 2017; 317: 197–208.
2. Chen XF, Gao W, Bi MS, Yang MR, Jiang HP, Zhang ZL, Zhang K. Experimental investigation on PMMA dust explosion venting at elevated static activation pressure. Powder Technol 2021; 392: 402–411.
3. Jiang HP, Bi MS, Gao ZH, Zhang ZL, Gao W. Effect of turbulence intensity on flame propagation and extinction limits of methane/coal dust explosions. Energy 2022; 239: 122246.
4. Yang MR, Jiang HP, Chen XF, Gao W. Characteristic evaluation of aluminum dust explosion venting with high static activation pressure. Process Saf Environ Prot 2021; 152: 83–96.
5. Gao W, Yu JL, Li J, Zhang Q, Xie QF, Zhang XY, Hu DP. Experimental investigation on micro- and nano-PMMA dust explosion venting at elevated static activation overpressures. Powder Technol 2016; 301: 713–722.
6. Yan XQ, Yu JL. Overpressure characteristics of aluminum dust explosion vented though a relief pipe. J Loss Prev Process Ind 2013; 26: 676–682.
7. Sazal KK, Jafar Z, Daniel E, Behdad M. Explosion severity of methane-coal dust hybrid mixtures in a ducted spherical vessel. Powder Technol 2018; 323: 95–102.
8. Yu MG, Li SS, Zhai F, Lou RY, Li HT, Zheng K. Effects of volatile contents on the deflagration characteristics and soot formation of methane/coal volatiles hybrid explosion using reaction kinetics simulation. Fuel 2022; 319: 123767.
9. Nie BS, Liu XF, Yang LL, Meng JQ, Li XC. Pore structure characterization of different rank coals using gas adsorption and scanning electron microscopy. Fuel 2015; 158: 908–917.
10. Li HY, Ogawa YJ, Shimada S. Mechanism of methane flow through sheared coals and its role on methane recovery. Fuel 2003; 82: 1271–1279.
11. Tan B, Shao ZZ, Xu B, Wei HY, Wang T. Analysis of explosion pressure and residual gas characteristics of micro-nano coal dust in confined space. J Loss Prev Process Ind. 2020; 64: 104056.
12. Sun S, Wang MY, Qiu YY, Gao KH. Study of flame propagation in an external space under vented explosion conditions. Energy 2019; 178: 186–194.

Chapter 6
Research on the Effects of Coal Dust Parameters on the Venting Characteristics of Gas/Dust Two-Phase Explosions

6.1 Experimental Apparatus and Procedures

The venting experiments of gas/dust explosions was conducted based on the apparatus shown in Fig. 6.1. The main structure included six subsystems and was clearly described in the previous reference [1]. The venting vessel was a cylindrical structure ($D \times H$: 318 mm × 400 mm) with stainless steel material and a venting pipe was installed horizontally in the center of wall surface ($L \times D$: 184 mm × 100 mm). Two optical glasses with 1.50 MPa design pressure were installed at the front and rear of venting vessel to realize the observation of flame. Two electrodes were installed at the center of vessel and the mixture was ignited by 8 kV high-voltage. Hartmann dust spraying device was installed at the center of bottom end flange to realize coal dust injection, and connected with the premixed tank by the one-way and solenoid valves. Simultaneously, the coal dust was injected and formed the dispersed coal dust cloud inside the venting vessel due to the vortex motion after the high-velocity airflow passed through the hemispherical nozzle. The high-frequency pressure transmitter (the frequency (f): 200 kHz; the range (R): 0–2.5 MPa) was adopted to record the pressures of vessel interior, venting pipe and the local positions (Distance from venting port (L) = 500 mm and 1000 mm). Furthermore, the development process of venting flame was captured by the high-speed camera (Photron V1212; f = 4000 fps). The temperature distribution was monitored by a high-speed infrared thermal imager (f = 1000 fps). The execution of function and collection of data were implemented by self-programming (A data acquisition card with a single channel of 2 M/s).

Four coal dust diameters (D_{50} = 14.41, 35.59, 76.66 and 141.40 μm) and four coal dust concentrations (q = 100, 200, 300 and 400 g/m^3) were adopted in the experiments. The rupture pressure of film (P_{stat}) was 0.32 MPa and the injection pressure of coal dust (P_{st}) was 0.40 MPa. The explosion intensity was the greatest under this condition. This indicates that the coal dust distribution was the most uniform and its concentration reached the optimal value inside the vessel under the action of gravity after dust spraying. The mixture gas of determined concentration

Z. Wang and X. Cao, *Mechanism of Composite Explosion Inhibition and Venting Protections*, https://doi.org/10.1007/978-981-95-4809-5_6

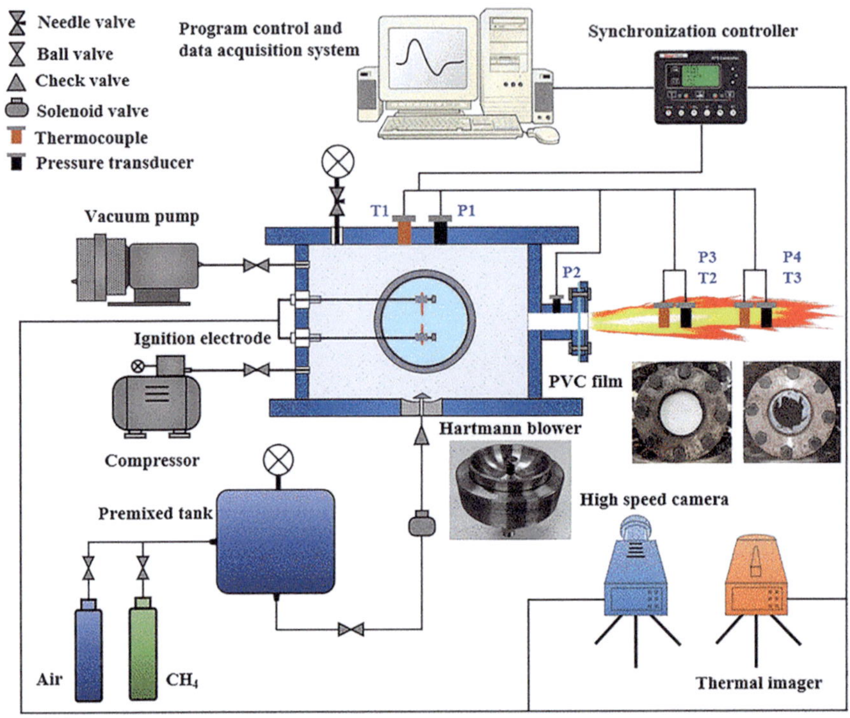

Fig. 6.1 Schematic diagram of experimental apparatus for researching the venting characteristics of gas/dust mixture explosions

was configured in the premixed tank to ensure the injection of coal dust and the concentration of methane inside the venting vessel ($c = 6\%$). And the component content and surface structure of coal dust were tested before the experiment by the industrial analysis and scanning electron microscope (ZEISS Sigma 300), as shown in Fig. 6.2. Its morphology and structure were smooth and uniform, and the diameter history presented the normal distribution (Mastersizer 2000). Besides, its weight loss rate was tested at high temperature (in Fig. 6.3). The coal dust at the heating rate of 10 K/min experienced the heat absorption, weight loss and stability stages. And it began to the pyrolysis after reaching 253 °C. TG history was rapidly decreased and the coal dust quality was gradually reduced within the range of 253–446 °C (the rapid weight loss stage). And the corresponding DTG history was also sharply increased and reached its peak value at 446 °C. Meanwhile, the DSC history of coal dust at this stage was also rapidly decreased, indicating that its internal pyrolysis reaction was enhanced and the heat release rate was increased. At 446 °C, the pyrolysis reaction was the most violent and the heat release was the highest. Within the range of 446–624 °C, the rate of weight loss was reduced (the slow weight loss stage). At this moment, the heat release was evidently decreased. After exceeding 624 °C, the TG-DTG-DSC histories tended to gentle and the coal dust reached the stable weight loss stage [2].

The venting vessel reached the determined vacuum degree before experiments. Meanwhile, the injection pressure of coal dust was calculated to ensure that a predetermined methane concentration was achieved after injection and the initial pressure reached 0.1 MPa. The dust injection time (t = 700 ms) and its setting time (t = 100 ms) were controlled by a program [3]. The explosion intensity was the greatest under this condition. This indicates that the coal dust distribution was the most uniform under the action of gravity after dust spraying. After ignition, the explosion flame of gas/dust mixture continued to develop and spread inside the vessel. As the

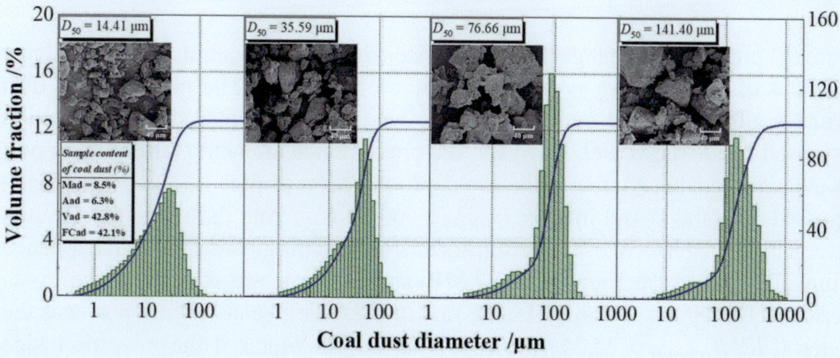

Fig. 6.2 Component content and surface structure of coal dust

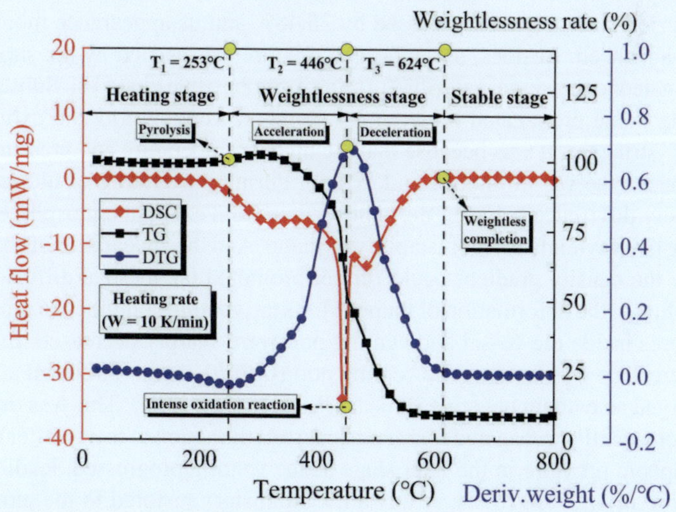

Fig. 6.3 Weight loss history of coal dust

pressure exceeded the film strength, the flames of gas/dust explosions were released to the vessel exterior, further causing the remarkable increases in the pressure and temperature. Besides, the deviation of ΔP_{max} at different positions was no more than 4.7% through the multiple experiments.

6.2 Results and Discussions

6.2.1 *Venting Characteristics*

Association Relationship Between Venting Parameters

Figure 6.4 presents the change of pressure rise with the propagation of venting flame both sides of the vessel ($q = 400$ g/m^3, $D_{50} = 35.59$ μm). The methane/coal dust mixture could be ignited and the pressure was sharply increased as the reaction progressed inside the vessel. However, the pressure rise moment at the venting port was evidently advanced due to the acceleration action of pipe ($t_{rise(2)} = 63.50$ ms) compared with the vessel interior ($t_{rise(1)} = 66.50$ ms). And the its rising rate was also greater than that of vessel interior. As the reaction progressed, the film happened rupture after the pressure reached 0.32 MPa and the flame was released to the vessel exterior at 108.00 ms. Meanwhile, the rate of pressure rise at venting port was the greatest ($(dP/dt)_{max(2)} = 24.34$ MPa s^{-1}). After the film ruptured, the pressures inside the vessel and venting port continued to rise due to the explosion reaction rate being significantly faster than the physical venting process. At 112.00 ms, the pressure of venting port reached the greatest ($\Delta P_{max(2)} = 0.42$ MPa). Compared with the vessel interior, $\Delta P_{max(1)}$ was increased by 35.48% and its appearance moment was evidently advanced. Besides, the unburned mixture was carried by the shock wave and propagated to the vessel exterior, forming a combustible cloud [4]. Subsequently, the venting flame propagated horizontally along the venting port and exhibited an "irregular" structure. It was because that the higher temperature and pressure differences between the vessel interior and exterior during the venting would result in a large density difference between the venting flame and external atmosphere, further promoting the Rayleigh Taylor instability of flame. And the appearance of shear layer caused by the density gradient could further promoted the thermal diffusion instability, leading to the deformation of flame [5]. As the venting continued to progressed, the pressures inside the vessel and venting port were sharply decreased. Especially the pressure drop was faster at the venting port ($(dP/dt)_{min(2)} = 55.92$ MPa s^{-1}) and even emerged an evident negative value ($\Delta P_{min} = -0.05$ MPa). This was due to that the flame constantly propagated forward because of the action of inertia after reaching the atmosphere pressure in the later stage as the venting progressed, leading to the sudden drop in pressure [6, 7]. Afterwards, the history restored to the atmosphere pressure.

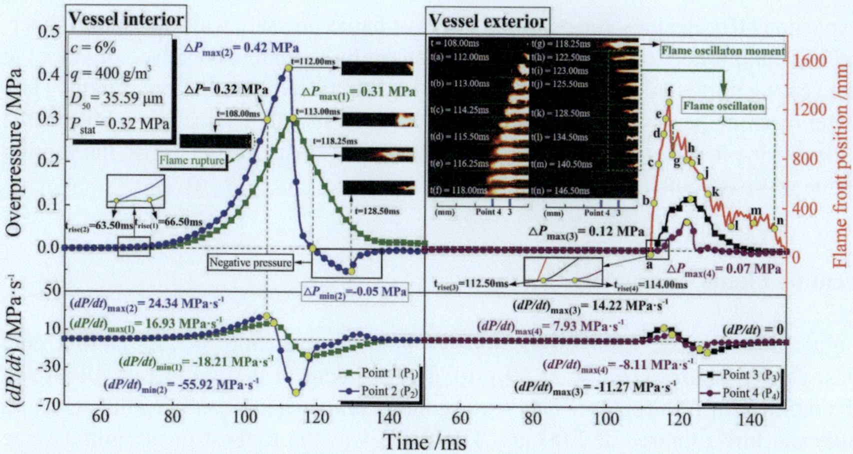

Fig. 6.4 Corresponding relationships between the pressure rise and flame propagation both sides of the vessel ($q = 400$ g/m^3, $D_{50} = 35.59$ μm)

The external pressure rise also showed a clear correspondence with the propagation of flame. The flame propagated horizontally and presented an "irregular" structure in the early stage of venting after the film ruptured. Thereafter, it was gradually transformed into a "mushroom" structure. The reason was due to the vortex structure resulted from the baroclinic effect [8]. Under the action of high pressure difference resistance, the nonparallel density and pressure gradients between the flow layers inside the venting flame would lead to the appearance of vortex structure, further resulting in an evident radial velocity component [9]. The pressure at 500 mm position distance from venting port started to rise at 112.50 ms, but the flame did not reach to this position (Point 3). It was because that the shock wave of gas/dust explosions was prior to flame propagation reaching this position. Due to the acceleration effect of venting pipe, the flame propagation velocity at the vessel exterior was obviously increased. It resulted in that the pressure at 1000 mm position distance from venting port began to rise at 112.50 ms (Point 4). However, its rise rate was obviously decreased and $\Delta P_{\max(4)}$ was decreased by 41.67% compared with Point 3. Simultaneously, the venting flame transformed from the "mushroom" to "spherical" structure. As the vortex motioned, its interior area was filled with the burned flame. And the unburned substance carried by the shock wave was ignited, further forming a "spherical" structure as the flame propagated. At 118.00 ms, the flame reached the furthest distance. Subsequently, the explosion reaction did not be sustained anymore with the consumption of unburnt substance and the venting flame started to extinguish, further forming a "strip" structure. At 118.25 ms, the venting flame emerged an evident oscillation phenomenon by the observation of flame front position. And its appearance moment was consistent with the negative pressure at the venting port. The phenomenon was attributed to the interaction between the negative pressure resulted from the venting inertia and positive pressure resulted from the external explosion [10]. Besides, the coupling effect of flame-pressure wave due to the superposition and reflection of pressure wave caused by the internal surface of vessel at the venting port would further promote the high-frequency oscillation of flame [11]. After reaching peak value, the pressure history was sharply reduced. In addition, the drop rate in pressure at Point 4 was faster than that of Point 3. Afterwards, the venting flame was extinguished completely.

Venting Flame Characteristics

Figure 6.5 illustrates the comparisons of flame propagation process at four coal dust concentrations ($D_{50} = 35.59$ μm). The moment in the figure was relative to the moment of film rupture. The venting flame propagated in an irregular structure after the film ruptured at 1.00 ms. The flame was the longest in the initial stage and presented a clear "mushroom" structure as its propagated as $q = 300$ g/m^3. Its surface area was decreased as the dust concentration changed. Meanwhile, the "mushroom" flame structure became less prominent. Especially the flames always propagated in a "strip" structure as $q = 100$ and 200 g/m^3. Thereafter, the flame at 300 g/m^3 dust concentration was gradually transformed into an "spherical" structure, which was more pronounced than at 400 g/m^3 dust concentration, and appeared an upward trend due to greater resistance and buoyancy. At 5.50 ms, the flame firstly reached the longest distance at 300 g/m^3 coal dust concentration ($L_{max} = 1281$ mm). Followed by 400 g/m^3 coal dust concentration ($L_{max} = 1163$ mm) at 6.00 ms. As $q = 100$ and 200 g/m^3, the flame reached the longest distance at 6.50 ms ($L_{max} =$ 983 and 1032 mm). According to the velocity history (in Fig. 6.6), v_{max} emerged a changing regular of firstly increasing and then reducing with the improve of coal dust concentration (from 336.3 to 413.5 and 352.6 m/s). The propagation of flame was the fastest as $q = 300$ g/m^3, indicating the highest venting intensity. Followed by 400 g/m^3 coal dust concentration. The explosion intensity at coal dust concentrations of 100 and 200 g/m^3 were the smallest. After propagating to the farthest distance, the venting flame began to extinguish in a swinging "strip" structure. At this stage, the venting flames at four coal dust concentrations all exhibited varying degrees of oscillation phenomenon. As the coal dust concentration improved, the oscillation extent appeared the same changing regular as the velocity history (in Fig. 6.6). As $q = 300$ g/m^3, the average amplitude of flame oscillation was the largest ($A_{avg} =$ 134.3 mm). Meanwhile, an evidently secondary venting flame with an "umbrella" structure also appeared at the venting port. As the change of coal dust concentration, A_{avg} was decreased successively and the flame did not appear the secondary venting phenomenon. This fully demonstrates that the secondary venting was related to the explosion intensity of vessel interior and was evidently affected by the coal dust concentration.

Figure 6.7 shows the comparisons of venting flame at four coal dust diameters ($q = 300$ g/m^3). At 1.00 ms, the flame propagated in an "irregular" structure along the horizontal direction after the film ruptured. It clear that the length of venting flame did not differ significantly. As the venting progressed ($t = 2.00$–3.00 ms), its

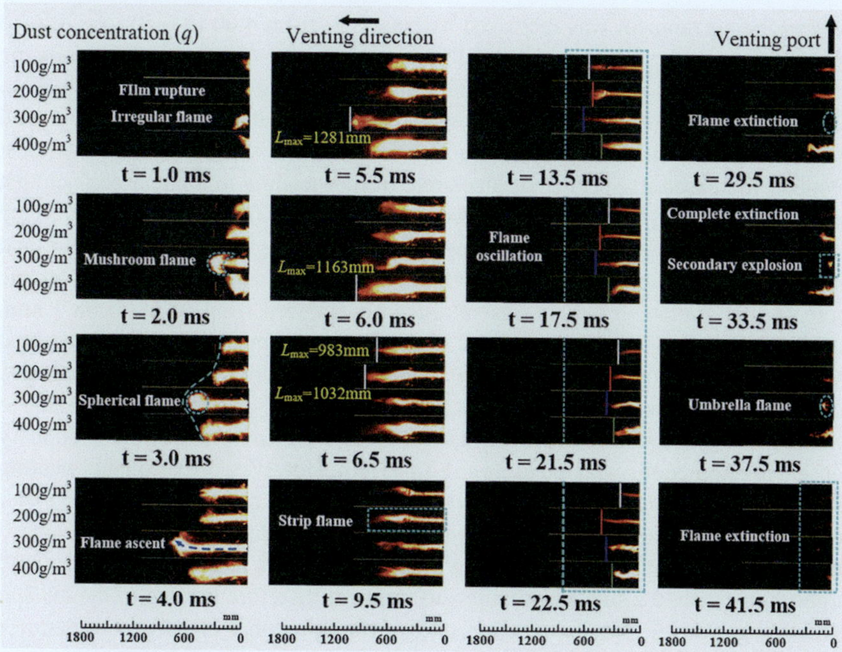

Fig. 6.5 Comparisons of venting flame under at four coal dust concentrations ($D_{50} = 35.59$ μm)

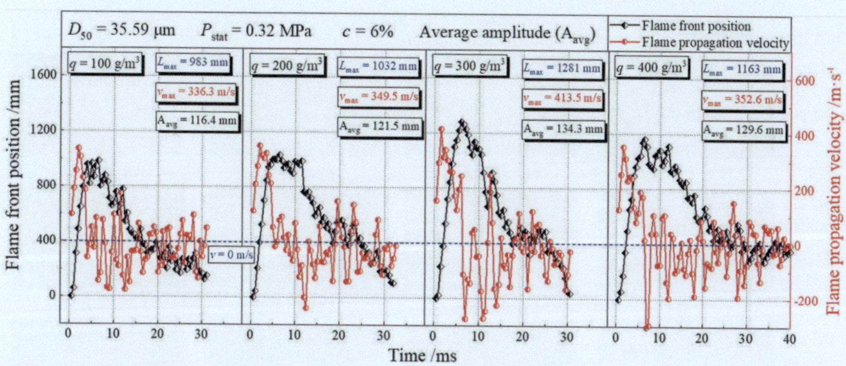

Fig. 6.6 Comparisons of velocity histories at four coal dust concentrations ($D_{50} = 35.59$ μm)

length was decreased successively as the coal dust diameter improved. As $D_{50} =$ 14.41 μm, the venting flame was slightly advanced and its surface area was larger. Under 35.59 μm coal dust diameter, the "mushroom" and "spherical" structures were more prominent during the venting. As the coal dust diameter continued to increase, the change in flame structure was not significant. Especially the flame did not appear the obvious "mushroom" structure as $D_{50} = 141.40$ μm. Meanwhile, the maximum venting length was evidently decreased from 1191 to 956 mm and its corresponding moment was increased successively from 4.00 to 6.50 ms with the increase in coal dust diameter. In addition, the velocity of venting flame was the fastest as $D_{50} = 14.41$ μm ($v_{max} = 363.9$ m/s), indicating the maximum explosion venting intensity, as can be seen in Fig. 6.8. The velocity history was decreased as the diameter continued to increase. This was because that the smaller coal dust diameter increased its specific surface area, further leading to a larger contact area with oxygen during the mixture explosion and causing a faster reaction rate. In the later stage of venting, the flame extinguished in a swinging "strip" structure and exhibited the varying degrees of oscillation affected by coal dust diameter. With the increase of the diameter, A_{avg} was significantly reduced from 187.8 to 98.7 mm. And the oscillation extent was related to the explosion intensity and was affected by the diameter of coal dust. It was enlarged with the increase of venting intensity. Afterwards, the unburned mixture was rapidly consumed during the venting due to the intense reaction of coal dust with smaller diameter. Hence, no obvious secondary venting phenomenon appeared as $D_{50} = 14.41$ μm. However, the secondary venting phenomenon appeared firstly under 141.40 μm coal dust diameter. As the diameter continued to increased, the appearance moment of secondary venting was advanced and the phenomenon was also more significant. It was due to that the coal dust with larger diameter had a smaller specific surface area, therefore leading to the evident reduce in gas/dust explosion reaction rates. A large number of unburned mixtures remaining inside the vessel was quickly released to the exterior, further producing a more significant secondary venting process [12].

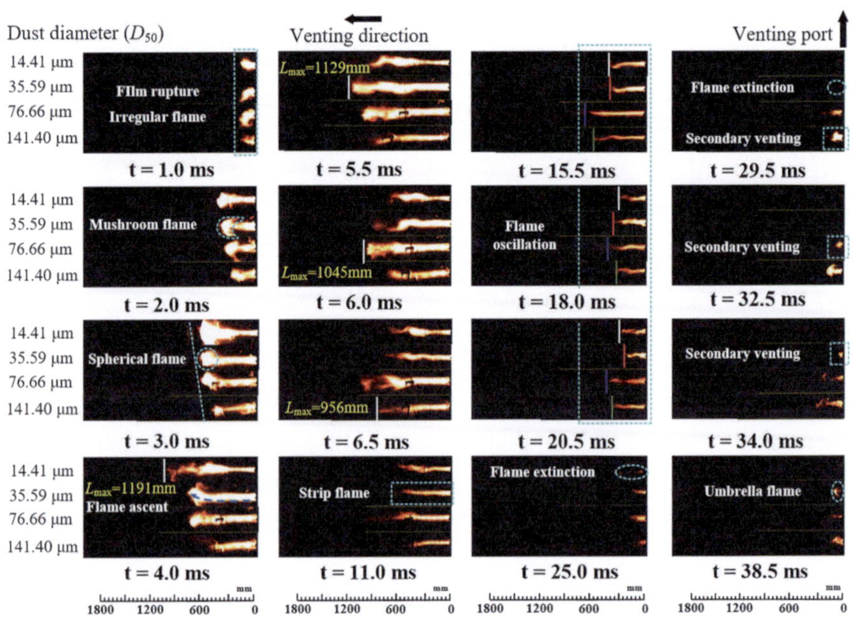

Fig. 6.7 Comparisons of venting flame at four coal dust diameters ($q = 300$ g/m^3)

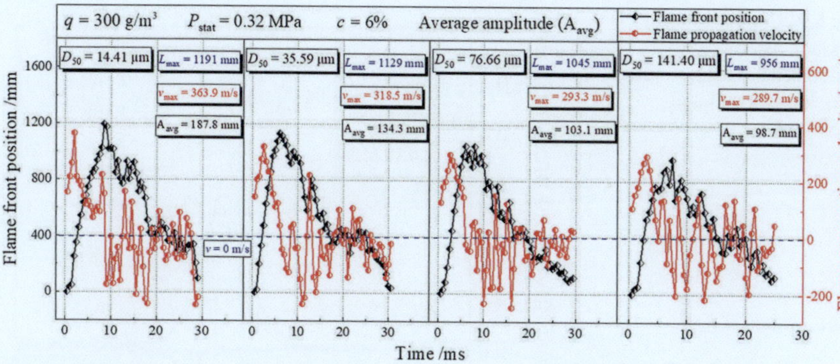

Fig. 6.8 Comparisons of velocity histories at four coal dust diameters ($q = 300$ g/m^3)

6.2.2 *Venting Pressure*

Figure 6.9 is the comparisons of pressure histories at different positions during the venting ($q = 300$ g/m^3, $D_{50} = 35.59$ μm). It clear that there was an evident difference in pressures at different positions. As the venting progressed, the pressure history from the vessel interior to exterior displayed a variation process of firstly increasing and then reducing. The pressure at the venting port was the greatest ($\Delta P_{\max(2)} =$ 0.45 MPa) and its corresponding moment was the smallest ($t_{\max(2)} = 109.50$ ms). The corresponding moment of pressure rise was evidently advanced and its rise rate was increased than that of interior, and the drop rate of pressure was also more remarkable. Besides, the pressure history emerged an evident negative value ($\Delta P_{\min}$ $= -0.06$ MPa). The pressure inside the vessel ($\Delta P_{\max(1)} = 0.28$ MPa) was higher than that of vessel exterior. With the increase of venting distance, the external pressure was gradually reduced. The pressures at Point 3 and 4 were decreased respectively by 62.8% and 74.4% compared with the venting port.

Influence of Coal Dust Concentration

The change in pressure distribution could be evidently affected by the coal dust parameter during the venting. Figure 6.10 illustrates the comparisons of pressure histories inside the venting pipe at four coal dust concentrations ($D_{50} = 35.59$ μm). The pressure showed a changing regular of firstly increasing and then reducing with the improve of coal dust concentration. There was an optimal coal dust concentration corresponding to the maximum venting intensity of mixture explosions. As q $= 300$ g/m^3, the reaction rates of mixture explosions were the most violent and the corresponding pressure was the greatest. It clear reflected in the evident reduction of rise moment and the increase of slope in pressure history. In particular, $\Delta P_{\max}$ was the greatest and the corresponding moment was advanced. As the coal dust concentration changed, the moment of pressure rise was obviously prolonged and its slope was

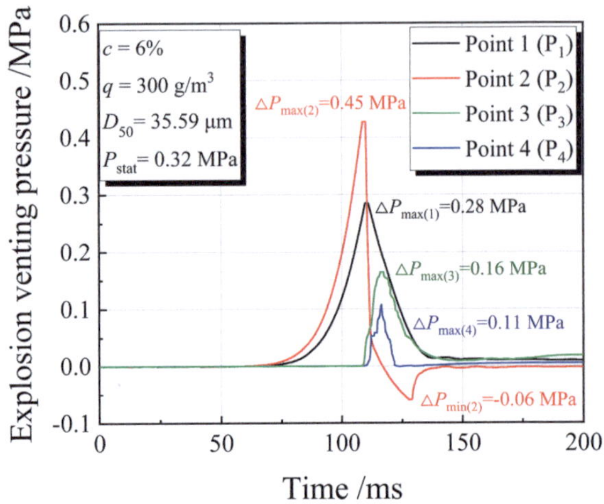

Fig. 6.9 Comparisons of pressure histories at four positions during the venting ($q = 300$ g/m^3, $D_{50} =$ 35.59 μm)

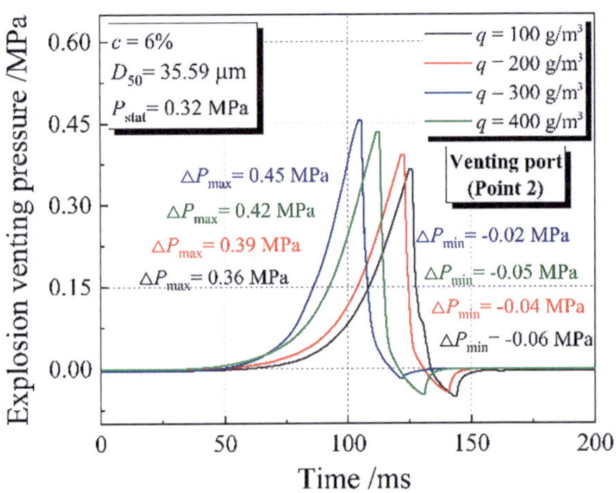

Fig. 6.10 Comparisons of pressure histories inside the venting pipe at four coal dust concentrations ($D_{50} =$ 35.59 μm)

also continuously decreased. Especially $\Delta P_{\max}$ was reduced by 6.67%, 13.33% and 20.00%, respectively. Besides, the appearance moment of negative pressure showed a changing regular of firstly decreasing and then increasing as the coal dust concentration improved. This fully demonstrates that the dust concentration could evidently affect the intensities of gas/dust explosions, thereby affecting its pressure distribution during the venting.

Influence of Coal Dust Diameter

Figure 6.11 is the comparisons of pressure histories inside the venting pipe at four coal dust diameters ($q = 300$ g/m^3). The venting pressure showed a decreasing variation

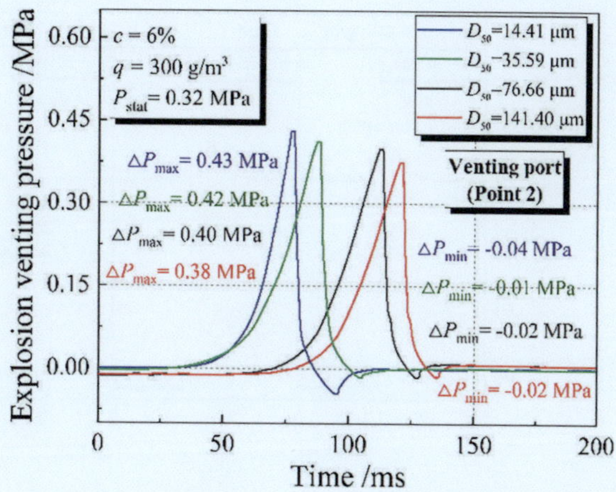

Fig. 6.11 Comparisons of pressure histories inside the venting pipe at four coal dust diameters ($q = 300$ g/m^3)

regular with the improve of coal dust diameter. The corresponding pressure rise moment was continuously delayed and its slope was decreased successively. As $D_{50} = 14.41$ μm, the pressure at the venting pipe was the greatest ($\Delta P_{max} = 0.43$ MPa) and the corresponding moment of ΔP_{max} was the smallest. As $D_{50} = 141.40$ μm, the pressure inside the venting pipe was the smallest ($\Delta P_{max} = 0.38$ MPa) and its corresponding moment was largest. The ΔP_{max} was reduced by 4.65, 6.98 and 15.56% as the coal dust diameter improved. The corresponding moment of negative pressure was also evidently delayed. It clearly explains that the diameter of coal dust had a dramatically impact on the venting pressure. The reason was that the decrease of diameter led to a continuous increase in the specific surface area and the obvious increase of contact area with oxygen per unit volume. Furthermore, the reaction rate of gas/dust explosions and its venting intensity all were enhanced.

6.2.3 Venting Pressure Rising Rate

The rate of pressure rise was an important parameter for the venting intensity. Figure 6.12 presents the comparisons of (dP/dt) histories at different positions during the venting ($q = 400$ g/m^3, $D_{50} = 35.59$ μm). It showed the same changing regular as the pressure history with the improve of venting distance. In particular, $(dP/dt)_{max}$ showed a variation regular of firstly increasing and then reducing, and the pressure change at venting port was most dramatically. Compared with the vessel interior ($(dP/dt)_{max(1)} = 14.46$ MPa/s), $(dP/dt)_{max(2)}$ was increased by 51.9% at the venting port due to the flame acceleration inside the venting pipe. And the pressure drop was also more significant because of the sudden expansion of venting area. As the venting distance continuously improved, the (dP/dt) was evidently decreased. $(dP/dt)_{max(3)}$ and $(dP/dt)_{max(4)}$ were reduced to 12.26 and 9.16 MPa/s. It was reduced by 44.17% and 58.29% respectively compared to the venting port. Compared with

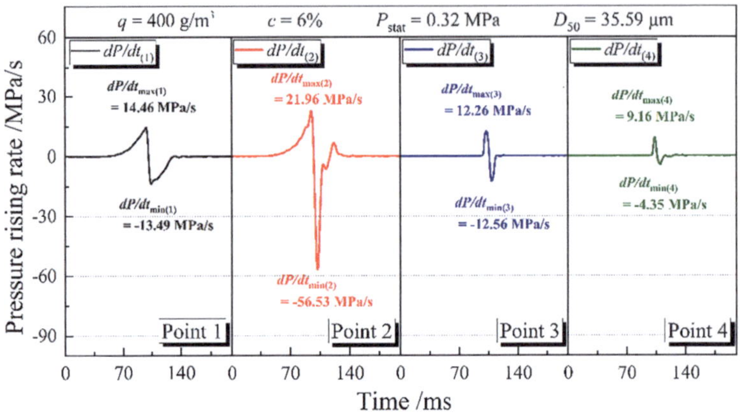

Fig. 6.12 Comparisons of (dP/dt) histories at different positions during the venting (q = 400 g/m^3, D_{50} = 35.59 μm)

the vessel interior, it was decreased by 13.8 and 36.7%. It was due to that the energies of shock wave and flame were sharply weakened during the propagation under the combined actions of heat dissipation and air resistance. Besides, $(dP/dt)_{min}$ also showed a variation regular of firstly increasing and then decreasing. Even the $(dP/dt)_{min}$ at the venting port reached −94.57 MPa/s.

Influence of Coal Dust Concentration

Figure 6.13 shows the comparisons of (dP/dt) histories at four coal dust concentrations (D_{50} = 35.59 μm). The history at the venting port presented a variation regular of firstly increasing and then reducing with the improve of coal dust concentration. The (dP/dt) was the greatest ($(dP/dt)_{max}$ = 20.57 MPa/s) as q = 300 g/m^3. Meanwhile, the rise and peak moments of history were also evidently advanced. As the coal dust concentration changed, (dP/dt) history was reduced. Especially for the absolute value of $(dP/dt)_{min}$. Besides, the (dP/dt) was the smallest at 100 g/m^3 coal dust concentration ($(dP/dt)_{max}$ = 14.01 MPa/s), and the rise and peak moments of (dP/dt) history were also evidently delayed. It fully demonstrates that the dust concentration could affect the reaction rates of mixture explosions, thereby affecting the rise rates of pressures at different positions during the venting.

Influence of Coal Dust Diameter

The diameter of coal dust was also an important factor for the rise rate of pressure during the venting (q = 300 g/m^3). (dP/dt) history at the venting port showed an obvious decreasing trend as the coal diameter improved, as can be seen in Fig. 6.14. The moment when (dP/dt) history started to increase was delayed and its peak

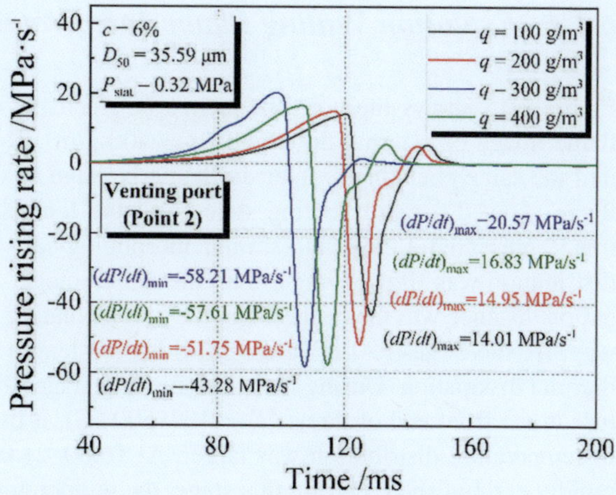

Fig. 6.13 Comparisons of (dP/dt) histories at the venting port under different coal dust concentrations ($D_{50} = 35.59$ μm)

moment was also prolonged. Especially $(dP/dt)_{max}$ was evidently decreased. As $D_{50} = 14.41$ μm, the (dP/dt) was the greatest ($(dP/dt)_{max} = 25.59$ MPa/s) and its peak value moment was the smallest (t = 72.60 ms). As the diameter improved, $(dP/dt)_{max}$ were respectively reduced by 29.46%, 35.56% and 39.70%. And their corresponding moments were evidently prolonged from 72.60 to 113.50 ms. Besides, the of $(dP/dt)_{min}$ and its appearance moment also exhibited a similar change regular. The absolute value of $(dP/dt)_{min}$ was decreased from 55.02 to 46.61 MPa/s. In addition, the corresponding moment was also increased from 80.40 to 123.20 ms. This further indicates that the dust diameter could evidently affect the reaction rates of mixture explosions, thereby also affecting the rise rate of pressure during the venting.

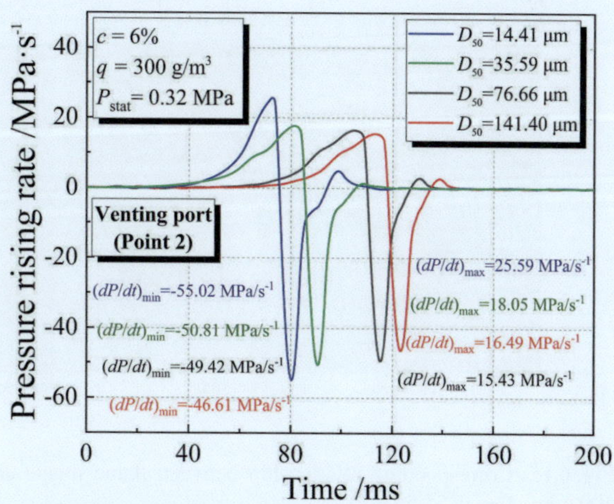

Fig. 6.14 Comparisons of (dP/dt) histories at venting port under different coal dust diameters ($q = 300$ g/m^3)

6.2.4 Explosion Venting Flame Temperature

Figure 6.15 shows the correspondence between the temperature distribution and flame image of external flow field (q = 400 g/m^3, D_{50} = 35.59 μm). Both exhibited the same changing regular. In the early stage of venting (t = 1.0–6.0 ms), the flame showed a rapid venting state. And the temperature distribution and flame image presented a clear and similar morphological structure, but the temperature distribution was slightly larger than the flame image. At 6.00 ms, the flame propagation distance was the farthest and the corresponding temperature distribution area was also the largest. Afterwards, its width and length continued to decrease due to thermal dissipation. During the extinguishing stage, there was still a high temperature in the front end of flame (T = 400–500 °C). It further resulted in that the area of temperature distribution was larger. At 16.00–21.00 ms, the venting flame was rapidly extinguished. During this stage, the temperature of venting flame front was decreased (below 500 °C). Simultaneously, the flame appeared the swing and formed a discontinuous temperature field structure and an oscillation phenomenon due to the thermal dynamic instability [10]. At 23.00 ms, the venting flame was completely extinguished and the corresponding temperature was reduced below 250 °C. After 25.00 ms, the flame appeared the secondary venting phenomenon with a "spherical" structure. It was because that the temperature inside the vessel continuously raised after the flame extinguished, causing the flame release again to the vessel exterior under the pressure difference and exhibiting a weak "spherical" structure. At this moment, the temperature of "spherical" flame center reached 400–500 °C. Afterwards, the flame was completely extinguished and the corresponding temperature contour disappeared after 27.00 ms.

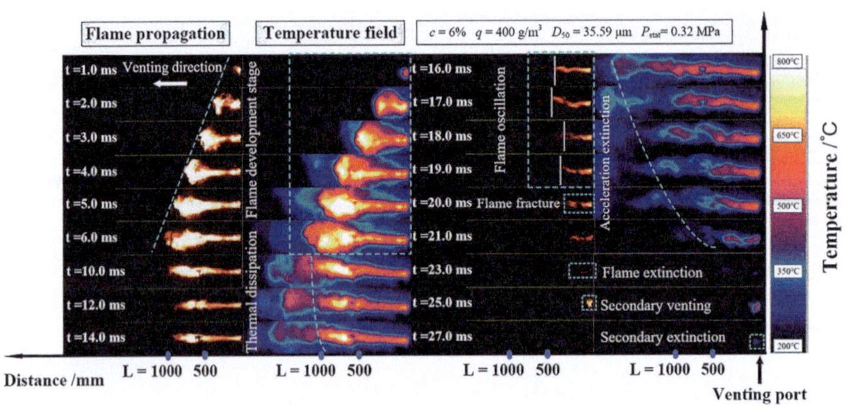

Fig. 6.15 Corresponding relationship between flame image and temperature distribution (q = 400 g/m^3, D_{50} = 35.59 μm)

Influence of Coal Dust Concentration

Figure 6.16 presents the correspondence of temperature distribution and its history along the venting direction at four coal dust concentrations ($D_{50} = 35.59$ μm). The temperature of flame front (T_{front}), the highest temperature (T_{max}), and venting port temperature (T_{port}) all emerged a variation regular of firstly increasing and then reducing under four coal dust concentrations. And the temperature at the corresponding position also showed the same trend of change as the coal dust concentration improved. T_{max} of venting flame did not appear on the flame front, but rather in the center of "spherical" flame near the front end. And the distance between the highest temperature and flame front continued to increase as the venting progressed. Under four working conditions, the venting flame temperature was the highest as $q = 300$ g/m^3, and T_{front}, T_{max} and T_{port} reached their maximum values at 4.00 ms ($T = 503$, 810 and 719 °C). And the corresponding flame width and surface area were also the largest. Follow by 400 g/m^3 coal dust concentrations. The corresponding temperature was reduced by 32.80%, 7.41% and 24.34%, respectively (t = 4.00 ms). As $q = 100$ g/m^3, the flame width and surface area were the smallest, and the temperature distribution always developed in a "strip" structure. T_{front} was lower than that of other three concentrations and the venting distance was the shortest. This fully demonstrates that the impact range of high temperature venting flame could be evidently affected by the concentration of coal dust. Especially for 300 g/m^3 coal dust concentration, the venting intensity was the most violent and the corresponding T_{front}, T_{max} and T_{port} all were the highest.

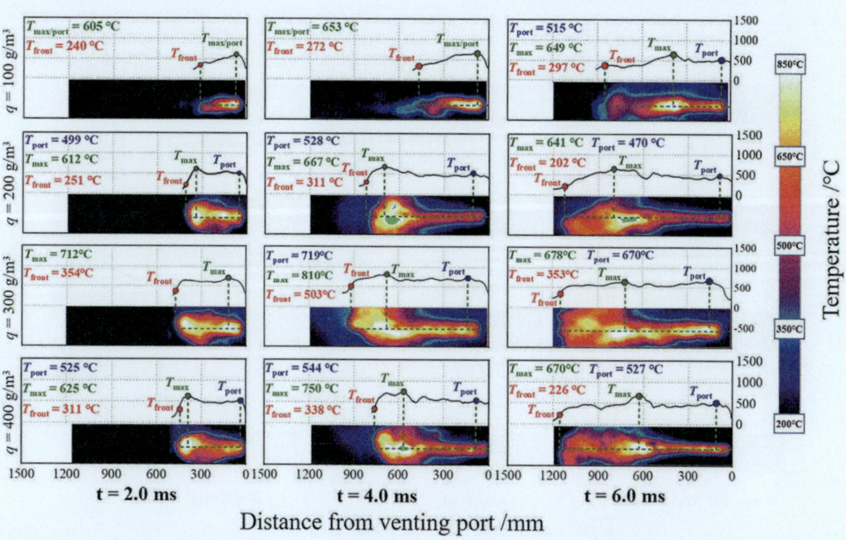

Fig. 6.16 Corresponding relationships between the temperature distribution and its history along the axis direction at four coal dust concentrations ($D_{50} = 35.59$ μm)

Influence of Coal Dust Diameter

Figure 6.17 presents the correspondence of temperature distribution and its history along the venting direction at four coal dust diameters ($q = 400$ g/m^3). As the coal dust diameter improved, the distribution range of temperature field was continuously decreased, and its width and length were also decreased. Especially the changes were more significant at 76.66 and 141.40 μm coal dust diameters. And T_{front}, T_{max} and T_{port} of venting flame all emerged a variation regular of firstly increasing and then reducing under four coal dust diameters. However, the temperature of venting flame presented a decreasing change regular as the diameter improved. Especially the changes of T_{front}, T_{max} and T_{port} were more significant. As the venting progressed, the distance between the highest temperature and flame front continued to increase. Especially this phenomenon became more obvious as the coal dust diameter increased. Besides, T_{front}, T_{max} and T_{port} reached the highest temperatures at 4.00 ms under 14.41 μm coal dust diameter (T = 512, 953 and 739 °C). It indicates that the venting intensities of gas/dust explosions were the greatest at this moment. Follow by 35.59 μm coal dust diameter ($T_{max} = 810$ °C). Compared with 14.41 μm coal dust diameter, T_{max} was decreased 15.01%. T_{max} were decreased to 691 and 625 °C under 76.66 and 141.40 μm coal dust diameters, and it was reduced by 27.49% and 34.42%, respectively. Besides, the distance between the flame front and highest temperature was gradually increased and the interval temperature was decreased. It also fully explains that the temperature distribution of external flow field was related to the intensities of mixture explosions and was evidently affected by coal dust diameter.

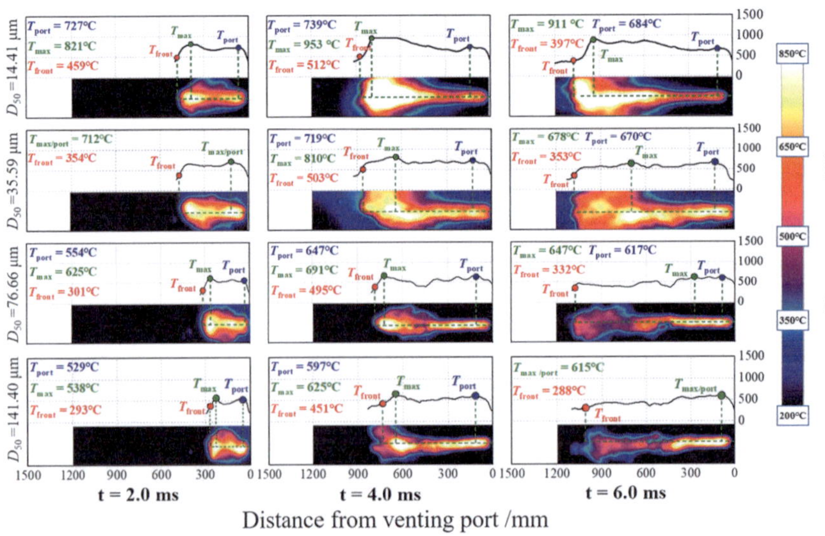

Fig. 6.17 Corresponding relationships between the temperature distribution and its history along the axis direction at four coal dust diameters ($q = 400$ g/m^3)

6.2.5 Mechanism Analysis

Figure 6.18 is a schematic diagram of gas/dust two-phase explosion venting processes. The coal dust was carried by the premixed gas and uniformly dispersed inside the vessel, forming an explosive gas/dust two phase mixtures. Under the action of 8 kV ignition energy, the continuous premixed gas was preferentially ignited. And the heat released by the gas explosion was quickly transferred to dispersed coal dust by the methods of convection and thermal radiation [13]. Subsequently, it was continuously transferred from the external surface to its interior, leading to the rapid pyrolysis and precipitation of volatile components in the coal dust [14]. The large molecule combustible substance was fractured into the small molecule gaseous (CH_4, CO_2, O_2, N_2, CO, etc.) and liquid products (hydrocarbons), resulted in the continuous expansion and increase of coal dust volume [15]. After reaching a certain pressure, volatile substances were continuously released from its surface, and then forming the pore coal dust structure. The released volatile gases would greatly improve the explosion reaction rate and the heat exchange between the gas and solid two phases. It further intensified the reaction rates of coal dust pyrolysis and residual coke multi-phase substance ($C + H_2O \rightarrow CO + H_2$, $C + CO_2 \rightarrow 2CO$, $2C + O_2 \rightarrow 2CO$, $C + O_2 \rightarrow CO_2$). After the mixture was ignited at the center of vessel, the shock wave propagated in the front of flame and was enhanced due to the acceleration of flame. The film ruptured as the pressure at venting port exceeded its strength. The unburned mixture could be carried by the precursor shock wave and propagated to the exterior of vessel, forming a combustible cloud near the venting port [16]. After the flame front contacted with an external combustible cloud, unburned substance including to the coal dust and premixed gas would happen the violent explosion reaction, forming a bright flame outside the venting port. Besides, the coupling process between the flame and shock wave would be further enhanced during the gas/dust explosion venting due to the interaction and superposition between the shock wave and the reflected wave resulted from the obstruction of inner wall surface.

6.3 Conclusions

The impact of coal dust parameters on the venting process of gas/dust explosions was conducted. The association relationship between the explosion venting parameters both sides of the vessel were analyzed, and the influence regular and mechanism of coal dust parameters on the dynamics characteristics of gas/dust two-phase explosion venting were also revealed. Conclusion are as follows:

First, a clear correspondence between the explosion parameters both sides of the vessel during the venting was presented. The venting flame experienced a development process from the "irregular" to "mushroom", "spherical" and "strip" structures, and was evidently affected by coal dust parameters. With the improve of coal dust concentration, the venting distance emerged a variation regular of firstly increasing

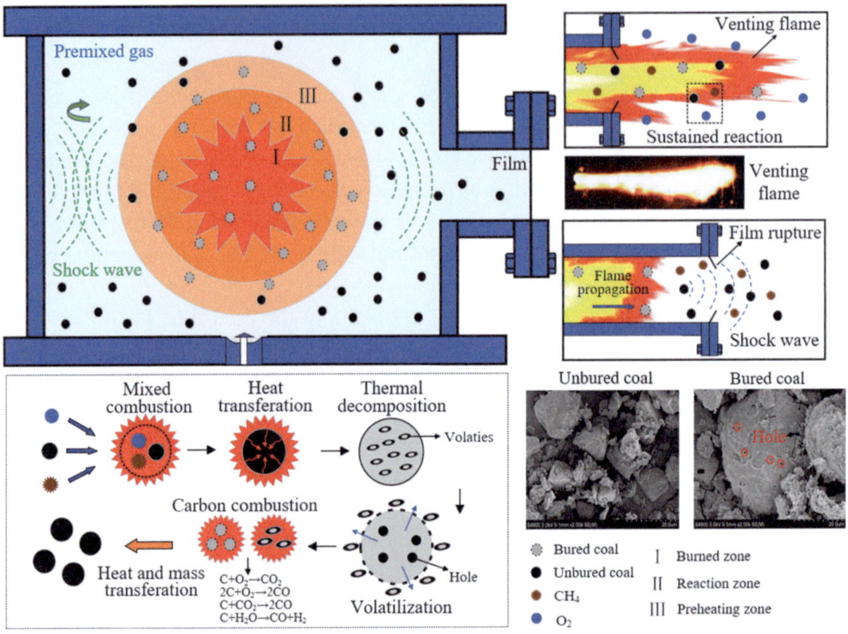

Fig. 6.18 Schematic diagram of gas/dust explosion venting process

and then reducing. And it was the longest as $q = 300$ g/m^3. Simultaneously, the velocity history also exhibited the same change regular. As the increase of its diameter, the length and velocity of venting flame all were decreased. Thereafter, the flame emerged the varying degrees of oscillation with the change of coal dust parameters. The appearance moment of oscillation corresponded to the negative pressure. As the coal dust concentration improved, the oscillation extent was similar to the change in the velocity history. As $q = 300$ g/m^3, A_{avg} was the greatest. With increasing coal dust diameter, A_{avg} was reduced. Besides, an obvious secondary venting flame emerged, and it was related to the explosion intensity and affected by coal dust parameters. The appearance moment of secondary venting was advanced and it was also more significant with the coal dust diameter improved.

Second, the pressure at different positions existed an obvious difference during the venting. The pressure and its rise rate all showed a variation regular of firstly increasing and then reducing from the vessel interior to its exterior. The pressure of venting port was the greatest. The external pressure was evidently reduced with the venting distance enlarged. And its distribution could be evidently affected by coal dust parameters. As the coal dust concentration improved, the pressure and its rise rate also appeared a changing regular of firstly increasing and then reducing. As $q = 300$ g/m^3, the venting rate was the most violent and its corresponding pressure was the highest. The negative pressure of venting port showed an opposite change trend. However, the pressure showed a decreasing trend as the diameter improved. The negative pressure also showed a delayed change trend.

Third, the flame image and temperature field distribution presented a clear correspondence in the acceleration venting stage. Due to the heat dissipation, the flame was continuously extinguished and its length was decreased. However, there was still a high temperature in front of the extinguished flame, resulting a larger temperature contour than the flame image. Due to the thermal dynamic instability, the flame appeared swing and formed a discontinuous temperature field structure. As the coal dust concentration increased, T_{front}, $T_{\max}$ and T_{port} all emerged the changing regular of firstly increasing and then reducing. Meanwhile, the corresponding flame width and surface area also presented the same change. The temperature distribution continued to decrease as the diameter reduced, and T_{front}, $T_{\max}$ and T_{port} all presented the decreasing trend. $T_{\max}$ appeared at the center of "spherical" flame near the front. As the venting progressed, the distance between the highest temperature and the flame front continued to increase. The temperature distribution was related to the intensities of mixture explosions and affected by coal dust parameters.

To achieve the chemical equipment protection against gas/dust explosions, the following suggestions are proposed according to the research content: (1) Influences of dust parameters on the intensities of mixture explosions should be fully considered during venting protection design for dust enterprises. (2) To effectively reduce the damages of explosion shock wave and high-temperature venting flame to the external environment, the effective inhibition method should be taken in combination with explosion venting for gas/dust explosion protections. (3) The venting pipe could accelerate the flame propagation and increase the shock wave strength. For the practical engineering, its length should be minimized after meeting the normal operation condition to avoid the pressure exceeding the design pressure of equipment due to the acceleration effect of pipe.

References

1. Cao XY, Huang R, Lu YW, Wang ZR, Fan LT, Zhou YQ, Xu JJ. Experimental research on the venting characteristics of methane and coal dust two-phase system mixture explosions. Fuel 2023; 353: 129208.
2. Guo R, Li N, Zhang XY, Zhang YS, Chen QY, Zhao WB, Wang X, Li RZ. Suppression mechanism of micron/nano PMMA dust flame based on thermal analysis. Adv Powder Technol 2022; 33: 103848.
3. Chen XF, Gao W, Yang MR, Jiang HP, Zhang ZL, Zhang K. Experimental investigation on PMMA dust explosion venting at elevated static activation pressure. Powder Technol 2021; 392: 402–411.
4. Mike K, Andreas F, Gerold S, Natalie K, Simon J, Beatrice L. Medium-scale experiments on vented hydrogen deflagration. J Loss Prev Process Ind 2015; 36: 416–428.
5. Hu QR, Shen XY, Huang ZR, Qian XM, Jiang JH, Yuan MQ, Pang L. Shock wave dynamics and venting overpressure hazards induced by indoor premixed hydrogen/air explosion. Int J Hydrogen Energy 2023.
6. Wang CH, Guo J, Zhang K, Du SF, Chen H, Yang FQ. Experiments on duct-vented explosion of hydrogen-methane-air mixtures: Effects of equivalence ratio. Fuel 2022; 308: 122060.

7. Vladimir M, Dmitriy M, Jonathan P. The nature and large eddy simulation of coherent deflagrations in a vented enclosure-atmosphere system. Loss Prev Process Ind 2006; 19: 121–129.
8. Chen XF, Xu JT, Jiang HP, Gao W. Effects of ignition location on the characteristics of dust explosion venting at elevated static activation pressure. Powder Technol 2023; 420: 118408.
9. Cao Y, Gao KH, Li B, Xie LF, Pan XH. Influence of vent size and vent burst pressure on vented ethylene-air explosion: Experimental and numerical study. Process Saf Environ Prot 2023; 170: 297–309.
10. Hisken H, Enstad GA, Middha P, Wingerden K. Investigation of concentration effects on the flame acceleration in vented channels. J Loss Prev Process Ind 2015; 36: 447–459.
11. Wang Z, Zhang Z, Tian W, Wang Z. Coupling effect of side explosion vent and wire mesh on suppressing methane explosion in a closed duct. J Loss Prev Process Ind 2022; 76: 104738.
12. Lin, S, Liu, Z, Qian, J, Li X. Comparison on the explosivity of coal dust and of its explosion solid residues to assess the severity of re-explosion. Fuel 2019; 251: 438–446.
13. Sazal KK, Jafar Z, Daniel E, Behdad M. Explosion severity of methane-coal dust hybrid mixtures in a ducted spherical vessel. Powder Technol 2018; 323: 95–102.
14. Yu M, Li S, Zhai F, Lou R, Li H, Zheng K. Effects of volatile contents on the deflagration characteristics and soot formation of methane/coal volatiles hybrid explosion using reaction kinetics simulation. Fuel 2022; 319: 123767.
15. Nie BS, Liu XF, Yang LL, Meng JQ, Li XC. Pore structure characterization of different rank coals using gas adsorption and scanning electron microscopy. Fuel 2015; 158: 908–917.
16. Cao Y, Wang ZY, Zeng MY, Chen JP, Li B, Xie LF. Experimental and numerical study of the influence of vent burst pressure on venting characteristic of hydrogen-air explosion. Nucl Eng Des 2022; 394: 111844.

Chapter 7
Effect of Operation Condition on the Venting Characteristics of Gas/Dust Two-Phase System Explosions

7.1 Experimental Apparatus and Procedures

Figure 7.1 shows the schematic diagram of experimental equipment for studying the venting process of two-phase system explosions. And the detailed structure has been introduced in the previous reference [1]. The main body of vessel was a vertical column structure ($H \times D$: 400 mm × 318 mm). And it was made of the stainless steel and the design pressure was 5.0 MPa. The venting duct ($L \times D$: 180 mm × 100 mm) was installed horizontally at the center of venting vessel side wall. The film was install at the venting port and compacted with the flange. Two ignition electrodes were placed in the center of vessel and the ignition voltage was 8 kV to realize the ignition of gas/dust mixture. The methane and air were mixed inside the premixed tank and the dust injection pressure was changed by controlling the high pressure gas.

The Hartmann dust spraying apparatus was adopted to realize raising process of coal dust. Simultaneously, it was installed at the center of vessel bottom end by threaded method and connected with premixed tank through an electromagnetic and one-way valves. The coal dust in the groove could be raised to form a uniform dust cloud due to the blocked by the hemispherical nozzle cover as the high-velocity airflow passed through it from the bottom tube bundle and the formed a vortex motion. Four high-frequency pressure transmitters (Response frequency (f): 50 kHz; Range (R): 0–2.5 MPa) were placed at the centers of vessel top end and venting duct, and the horizontal positions of 500 and 1000 mm far away the venting port (L) to record the changes of local pressures. Additionally, the propagation characteristics of venting flame was collected by a high-speed camera (Photron V1212) under different operation conditions ($f = 4000$ fps). And the temperature distribution was instantaneously record by a high-speed infrared thermal imager ($f = 1000$ fps). To ensure the synchronous acquisition of flame propagation and pressure data during the experiment, the program control system was compiled to complete synchronous control. During the experiment, the dust spraying system, ignition system, data acquisition

Z. Wang and X. Cao, *Mechanism of Composite Explosion Inhibition and Venting Protections*, https://doi.org/10.1007/978-981-95-4809-5_7

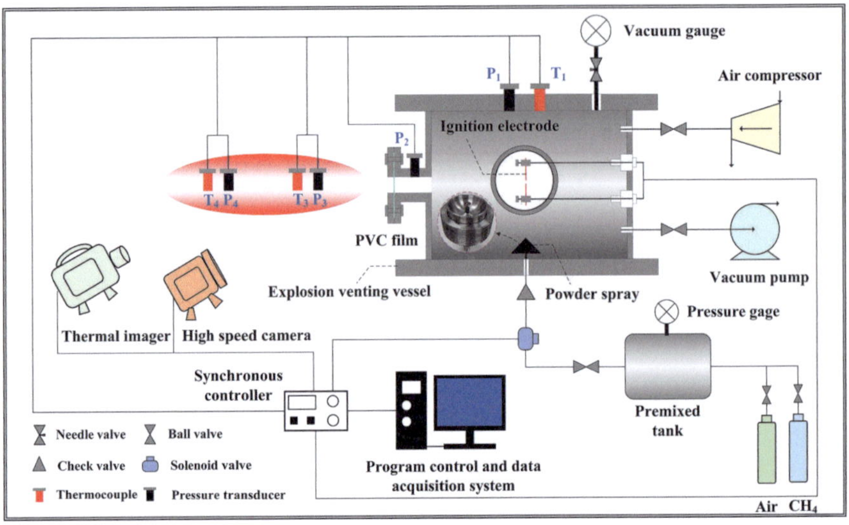

Fig. 7.1 Structure diagram of explosion venting experiment apparatus

system (include the pressure, flame image and temperature) were controlled by a program control and data acquisition system. And their order and interval time were also determined. By comparing the moment between the propagation of venting flame and the rise of pressure history after ignition, their corresponding relationship was accurately analyzed.

The physical parameter of coal dust was tested before the experiment (seen in Fig. 7.2). The diameter of coal dust was determined by laser particle size analyzer (Type: Mastersizer 2000; $D_{50} = 35.59$ μm) and it was placed in the dust groove. The content of component in coal dust was tested by the industrial analysis. Moreover, its surface microstructure was observed by the scanning electron microscope (ZEISS Sigma300). It clear that the morphology of coal dust was uniform state and its diameter history shown a normal distribution. Besides, its thermal stability and pyrolysis were also measured as the temperature increased. At a heating rate of 10 K/min, the coal dust underwent three stages (include the heating, weight loss and stability three stages). And it appeared the pyrolysis at 293 °C. Thereafter, the acceleration weight loss began and reached its peak value at 578 °C. The corresponding DTG history reached its peak and then weightlessness entered the deceleration stage. As the temperature raised to 718 °C, the weight loss entered the stable stage.

Four film rupture pressures (The static activation overpressure (P_{stat}) = 0.16, 0.24, 0.32 and 0.40 MPa) and four initial pressures ($P_{init} = -0.01, 0.00, 0.01$ and 0.02 MPa) were selected during the experiment. Different rupture pressures were achieved by changing the layer number of the venting film during the experiment. The vacuum degree inside the venting vessel was realized by the vacuum pump. By calculation, the concentration and pressure of the premixed gas in the premixed tank were determined. The premixed gas would carry the coal dust into the venting vessel after the solenoid

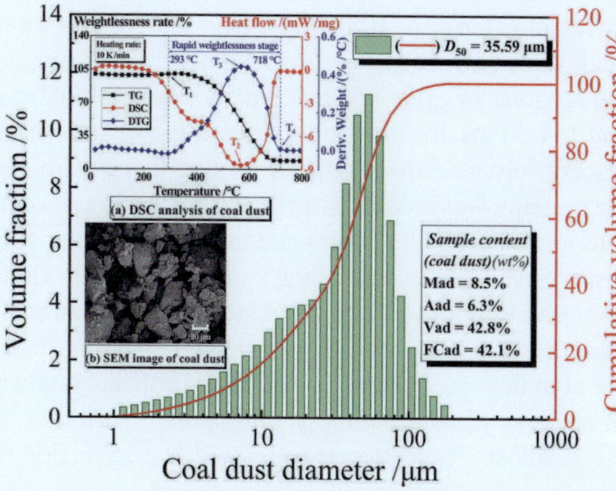

Fig. 7.2 Analysis of coal dust physical parameters

valve was opened, and a determined dust concentration ($q = 300$ g/m^3) and premixed gas concentration (c = 6%) inside the vessel were evenly formed. And the initial pressure was determined at the ignition moment. Through comparative experiments, the dust spraying pressure and time were determined (0.40 MPa and 700 ms). The mixture of gas and dust was ignited after the solenoid valve was closed for 100 ms. Then the flame of vessel interior continuously spread. Afterwards, it was released to the vessel exterior as the venting pressure exceeded the film rupture pressure, leading to the great increases in temperature and pressure of external environment. To ensure the accuracy of the experiment, each operation condition was repeated 4 times. By analyzing experiment data, the deviations of pressure peak at four positions were not more than 5.6%.

7.2 Results and Discussions

7.2.1 Explosion Venting Characteristics

Characteristic Analysis of Explosion Venting

Figure 7.3 is the correspondence between the venting flame propagation and pressure rise of vessel interior and exterior at atmospheric pressure ($P_{\text{stat}} = 0.40$ MPa). The mixture was ignited after the gas and dust were uniformly dispersed inside the vessel, and the flame spread and propagated toward to the wall surface, further resulting in a rapid pressure rise. Due to the acceleration effect of duct on the venting flame, the pressure was evidently increased and its rise moment was slightly advanced compared with the vessel interior. As the venting flame developed, the pressure of vessel interior was obviously increased. Afterwards, the venting film ruptured after the pressure at the venting port reached 0.40 MPa (film rupture pressure) and the flame was released to vessel exterior at 105.75 ms. And the corresponding (dP/dt) history reached its peak value (($(dP/dt)_{max}$ = 29.33 MPa s^{-1}). The pressure continued to increase with the progress of explosion reaction, but its rise rate began to decrease. As t = 110.75 and 111.00 ms, the pressures of venting port and vessel interior reached their peaks, respectively. And the lengths of venting flame were 1208 and 1314 mm. Afterwards, the pressure was reduced with the extinguishing of venting flame. Compared with the interior, the drop rate of pressure inside the venting duct was greater and the pressure history even presented a larger negative value. This is because that the pressure of vessel interior reached a balance with the external environment pressure as the flame extinguished in the later stage of venting. However, the flow field still had the high flow inertia and continued to propagate outward, resulting in the formation of negative pressure at the venting port [2]. At 127.00 ms, the negative value was the smallest. Thereafter, the pressure history inside the vessel reached the normal pressure after the flame was extinguished.

The flame in an irregular structure propagated horizontally along the venting port after the film ruptured. The flame was acceleratedly released due to the high pressure difference. Affected by air resistance, the surface area of flame was continuously enlarged and formed a "mushroom" structure. It was attributed to the vortex structure formed by the baroclinic effect. And the vortex structure would appear in the oblique area due to the nonparallel pressure and density gradients between the flow layers inside the venting flame. With the acceleration propagation of flame in the early stage, the surface area of flame front was continuously increased and its propagation velocity appeared the radial component under actions of high pressure difference and resistance. As the vortex developed, the flame continued to expand outward and ignited the forward unburnt combustible materials as its spread, further inducing the

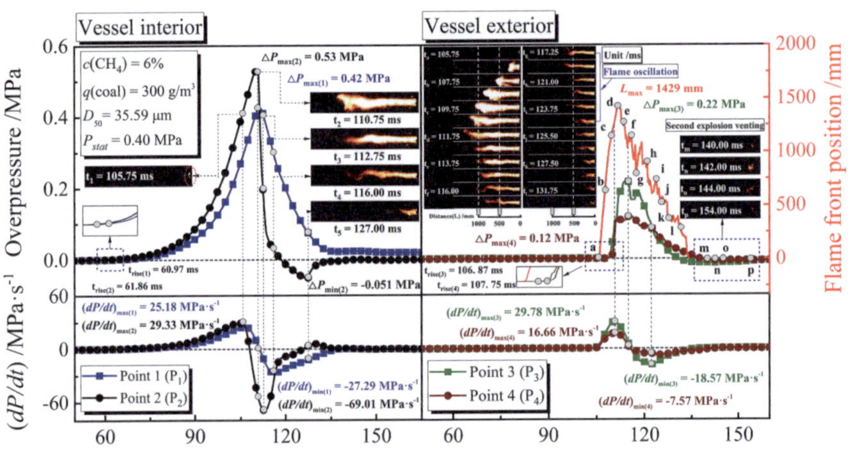

Fig. 7.3 Corresponding relationship between the venting flame propagation and pressure rise at atmospheric pressure (P_{stat} = 0.40 MPa)

flame front expansion and forming the mushroom cloud-like/ellipsoidal structures and large-scale vortex structures under the actions of diffusional-thermal instability, Kelvin–Helmholtz instability [3].

At 106.87 ms, the venting flame did not yet reach the P_3 monitoring point, but the pressure history started to rise under the action of shock wave. Compared with the vessel interior, the rise moment of pressure history outside the vessel was prolonged. Nevertheless, its rise rate was obviously increased. The rise rate of pressure history at point P_4 also was sharply increased due to the high velocity propagation of venting flame, but its rise moment presented a slight delay (t = 107.75 ms). Meanwhile, the pressure drop was obviously increased as the venting distance enhanced due to the rapid heat dissipation. Under the action of buoyancy, the venting flame tilted upwards. The maximum length appeared at 111.75 ms, but the pressure did not reach its peak. Compared with point P_4, the pressure rise rate at point P_3 was greater. Afterwards, the vortex motion of flame was decreased and disappeared eventually with the attenuation of venting flame. Thereafter, the flame near the outer layer continued to disappear and propagated in a "strip" structure as the heat continued to dissipate. The pressure histories at two external positions reached their peaks at 116.90 and 117.44 ms ($\Delta P_{max(3)} = 0.22$ MPa; $\Delta P_{max(4)} = 0.12$ MPa). Compared with vessel interior, the exterior pressure was evidently reduced and it was further decreased as the venting distance increased. Besides, the flame appeared an evidently oscillation during the flame extinguishing stage. This is because that the reflection and superposition of pressure wave caused by venting port front and rear wall surfaces, resulting in a coupling effect of flame-pressure wave and further producing a high-frequency oscillation of flame along the venting direction [4]. At t = 140.00–154.00 ms, a secondary venting appeared near the venting port. Namely the venting appeared again after the flame was extinguished. This is due to that the negative pressure action at the venting port caused the reflux of fresh air. Afterwards, it was ignited by the high temperature gas after the mixture concentration inside the vessel reached the explosion limit, further appearing a secondary venting phenomenon [5]. The secondary venting phenomenon was the reappearance of the venting flame after it disappeared from the outside of the vessel. As the venting flame extinguished, the pressure histories at points P_3 and P_4 reached the normal pressure.

Venting Flame Analysis

Figure 7.4 is the comparisons of venting flame under different film rupture pressures. Combined with Fig. 7.5, it clear that the flame appeared instability at the film rupture moment and propagated in an irregular structure. As the film rupture pressure enhanced, the length of venting flame was enlarged at the same moment. As $P_{stat} \geq$ 0.24 MPa, the flame all exhibited a "mushroom" structure and it became more prominent as the film rupture pressure increased. As $P_{stat} = 0.16$ MPa, the venting flame always propagated in a small "strip" structure and its venting length was the shortest. The flame gradually developed from the "mushroom" to "spherical" structure as the venting progressed. And the formation moment of "spherical" flame was advanced as the rupture pressure increased. At this moment, the farthest venting distance was increased sequentially from 795 to 1429 mm and its corresponding moment was decreased. Thereafter, the venting flame started to extinguish and its structure gradually changed from the "spherical" to "strip" structure, accompanied by a decrease in the oscillation flame. Meanwhile, the flame all appeared the oscillation phenomenon during the extinguishing stage. And the oscillation was the most significantly as P_{stat} = 0.40 MPa and its oscillation amplitude was obviously increased. This is because that a stronger coupling effect of flame-pressure wave was enhanced as the film rupture pressure increased, further affecting the oscillation extent of external flame. Besides, the flame all appeared the secondary venting process at 37.00 ms as P_{stat} ≥ 0.32 MPa. Especially, the flame appeared the third venting process at 48.00 ms under 0.40 MPa film rupture pressure. This is because that the increase of film rupture pressure led to an enhancement in the reflux rate of exterior gas, thereby promoting the occurrence of third venting process.

Figure 7.6 shows the comparisons of venting flame velocity histories under different film rupture pressures. As the film rupture pressure increased, the velocity history was markedly increased in the early stage of venting. In particular, v_{max} was increased ceaselessly from 275.5 to 393.4 m/s and its appearance moment was obviously decreased from 1.50 to 0.50 ms. It indicates that the enhancement of film rupture pressure could increase the propagation velocity of venting flame. Simultaneously, the flame exhibited an obvious oscillation phenomenon and the average amplitude (A_{avg}) was increased as the film rupture pressure enhanced from 134 to 147 mm. And the maximum venting length of flame also was increased. Figure 7.7 presents the comparisons of velocity histories under different initial pressures

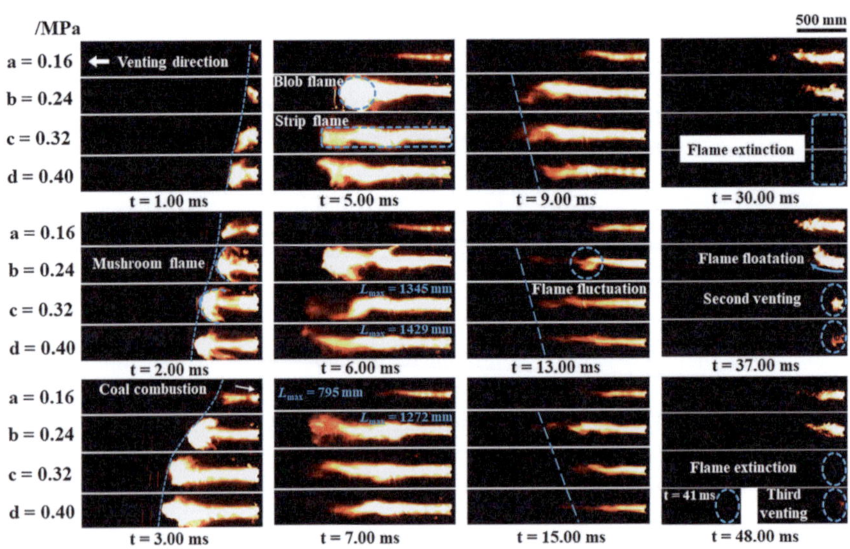

Fig. 7.4 Comparisons of venting flame propagation processes under different film rupture pressures

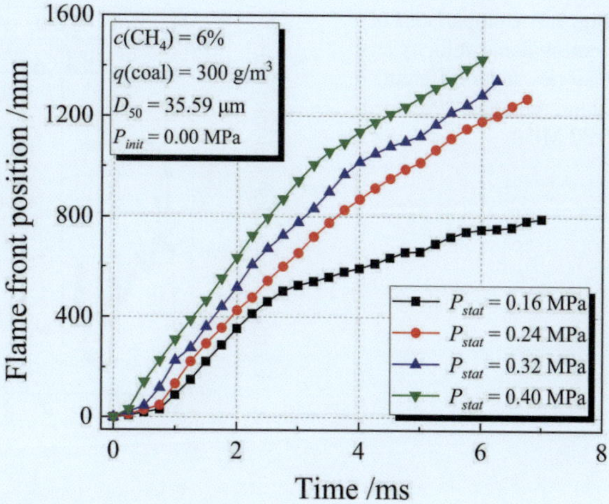

Fig. 7.5 Comparisons of venting flame front positions under different film rupture pressures

($P_{\text{stat}} = 0.32$ MPa). It can be seen that the velocity history was evidently increased as the initial pressure raised. During the venting initial stage, the velocity also appeared an obvious increase trend. In particular, v_{max} was increased from 313.2 to 388.7 m/s. In addition, the A_{avg} was increased from 141 to 152 mm as the initial pressure enhanced. And the maximum venting length of flame also was increased from 1295 to 1405 mm. This indicates that the explosion intensity of vessel interior could be prominently affected by the initial pressure, further affecting the propagation velocity of venting flame.

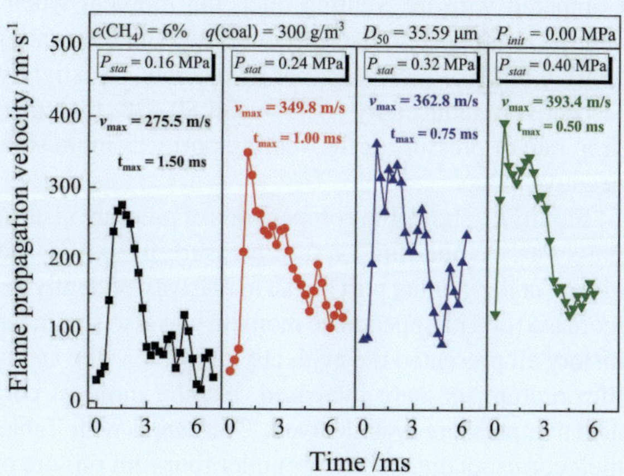

Fig. 7.6 Comparisons of venting flame velocity histories under different film rupture pressures

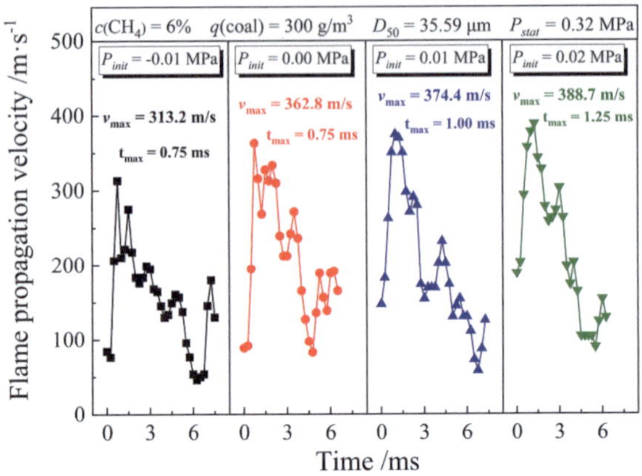

Fig. 7.7 Comparisons of venting flame velocity histories under different initial pressures ($P_{\text{stat}} = 0.32$ MPa)

7.2.2 *Explosion Venting Pressure*

Effect of Film Rupture Pressure

Figure 7.8 is the comparisons of venting pressure histories at different positions ($P_{\text{stat}} = 0.32$ MPa). It clear that the pressure histories presented an evident difference during the venting. The moment for the pressure rise started to rise at the venting port was slightly advanced compared with vessel interior. Due to the acceleration effect of venting duct on the flame, the corresponding venting pressure and its rise rate were evidently increased. Especially, the increase of $\Delta P_{\max}$ was more significant. Compared with the venting duct, the moment when pressure history outside the venting port began to rise was delayed. And the venting pressure was reduced as the venting distance increased, but its rise rate was slightly increased. The pressure was respectively reduced by 65.22% and 80.43% than that of venting port. Besides, the drop rate of pressure at the venting port was more significant and presented a larger negative value.

Figure 7.9 shows the comparisons of pressure histories under different film rupture pressures. As the film rupture pressure increased, the moment when the pressure history at the venting port began to rise was evidently delayed. $\Delta P_{\max}$ was ceaselessly increased and its appearance moment was also significantly delayed. And the pressure history all presented the evidently negative value and it was slightly increased as the film rupture pressure increased, but the moment corresponding to the maximum negative pressure was delayed. Combined with Table 7.1, the pressure peaks and their corresponding moments under four film rupture pressures at different positions all presented the same change trend.

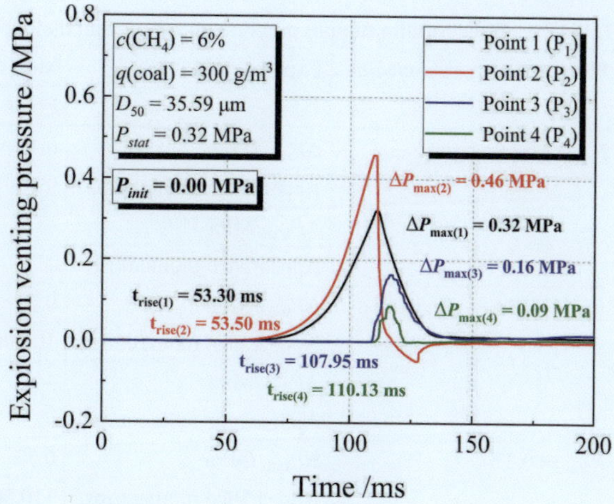

Fig. 7.8 Comparisons of venting pressure histories at different positions ($P_{stat} = 0.32$ MPa)

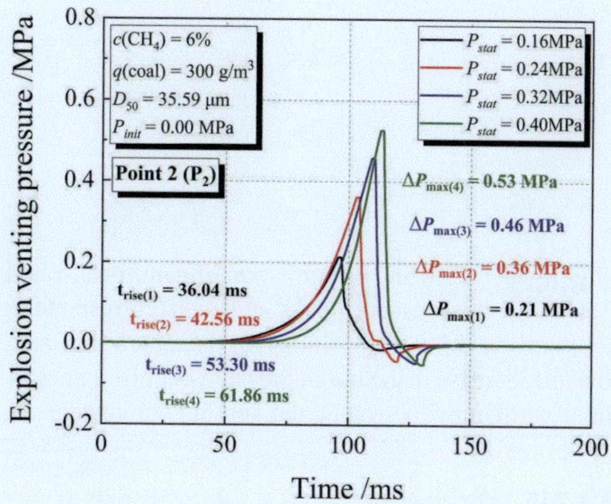

Fig. 7.9 Comparisons of venting pressure histories at venting port under different film rapture pressures

Effect of Initial Pressure

Figure 7.10 illustrates the comparisons of venting pressure histories at different positions ($P_{init} = 0.02$ MPa; $P_{stat} = 0.32$ MPa). The pressure history exhibited the same change pattern as the normal pressure condition as the venting position changed. But its rise moment was significantly advanced and the pressure inside the vessel was also enhanced. Compared with the vessel interior, the drop extent of exterior pressure was more significant. Figure 7.11 shows the effect of initial pressure on the pressure histories at the venting port ($P_{stat} = 0.32$ MPa). As the initial pressure increased, the moment when the pressure history started to rise was also obviously advanced, and its slope was successively increased. Meanwhile, ΔP_{max} was increased and its corresponding moment was decreased. Especially, the negative pressure was

Table 7.1 Effect of film rupture pressure on ΔP_{max} and their appearance moments

Film rupture pressure/MPa	Experimental data	Monitoring position			
		Point 1	Point 2	Point 3	Point 4
$P_{stat} = 0.16$	ΔP_{max}/MPa	0.15	0.21	0.09	0.03
	Appearance moment/ms	97.09	95.67	104.49	107.21
	ΔP_{min}/MPa	–	−0.02	–	–
	Appearance moment/ms	–	112.27	–	–
$P_{stat} = 0.24$	ΔP_{max}/MPa	0.26	0.36	0.12	0.07
	Appearance moment/ms	103.96	102.91	113.75	114.85
	ΔP_{min}/MPa	–	−0.04	–	–
	Appearance moment/ms	–	118.83	–	–
$P_{stat} = 0.32$	ΔP_{max}/MPa	0.33	0.46	0.16	0.09
	Appearance moment/ms	110.50	109.29	115.58	115.99
	ΔP_{min}/MPa	–	−0.05	–	–
	Appearance moment/ms	–	126.69	–	–
$P_{stat} = 0.40$	ΔP_{max}/MPa	0.42	0.53	0.22	0.12
	Appearance moment/ms	111.00	110.75	116.90	117.44
	ΔP_{min}/MPa	–	−0.05	−0.01	−0.01
	Appearance moment/ms	–	126.45	137.80	144.61

slightly increased and its corresponding moment was also decreased. Combined with Table 7.2, the pressure peaks and their corresponding moments under four initial pressures at different positions all presented the same change trend. This indicates that the increase in the initial pressure could enhance the explosion intensity of vessel interior, thereby increasing the venting parameters of gas/dust mixture explosions.

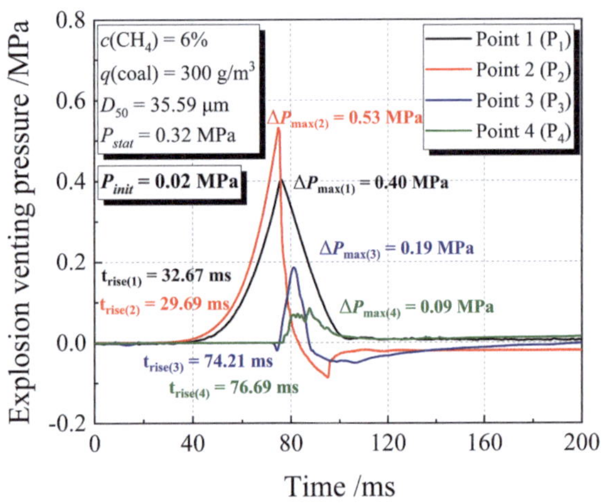

Fig. 7.10 Comparisons of venting pressure histories at different positions (P_{init} = 0.02 MPa; P_{stat} = 0.32 MPa)

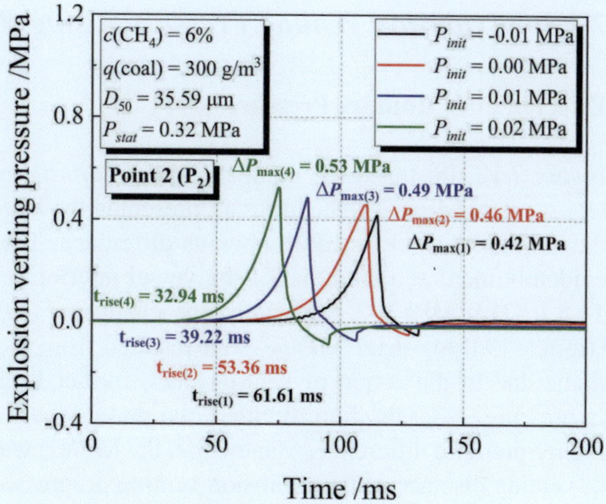

Fig. 7.11 Comparisons of venting pressure histories at venting port under different initial pressures ($P_{stat} = 0.32$ MPa)

Table 7.2 Effect of initial pressure on ΔP_{max} and their appearance moments

Initial pressure/MPa	Experimental data	Monitoring position			
		Point 1	Point 2	Point 3	Point 4
$P_{init} = -0.01$	ΔP_{max}/MPa	0.30	0.42	0.12	0.08
	Appearance moment/ms	115.38	114.67	120.58	119.62
	ΔP_{min}/MPa	–	−0.05	–	–
	Appearance moment/ms	–	131.17	–	–
$P_{init} = 0.00$	ΔP_{max}/MPa	0.32	0.46	0.16	0.09
	Appearance moment/ms	110.50	109.29	115.58	115.99
	ΔP_{min}/MPa	–	−0.05	–	–
	Appearance moment/ms	–	126.98	–	–
$P_{init} = 0.01$	ΔP_{max}/MPa	0.37	0.49	0.17	0.09
	Appearance moment/ms	86.75	86.62	93.33	93.25
	ΔP_{min}/MPa	–	−0.07	–	–
	Appearance moment/ms	–	105.85	–	–
$P_{init} = 0.02$	ΔP_{max}/MPa	0.40	0.53	0.19	0.09
	Appearance moment/ms	75.77	74.93	81.11	87.59
	ΔP_{min}/MPa	–	−0.09	−0.10	–
	Appearance moment/ms	–	94.75	106.90	–

7.2.3 Explosion Venting Pressure Rising Rate

Effect of Film Rupture Pressure

Figure 7.12 illustrates the comparisons of (dP/dt) histories at different positions ($P_{stat} = 0.32$ MPa). The histories all presented the largest and smallest peaks during the venting, but it existed an obvious difference. The (dP/dt) of venting port was evidently increased than that of the vessel interior. $(dP/dt)_{max}$ was increased from 14.5 to 21.9 MPa s^{-1}. Especially, the change of $(dP/dt)_{min}$ was more significant (from −19.0 to −61.1 MPa s^{-1}). It resulted from the acceleration propagation of flame due to the action of venting duct, further leading to a significant increase in pressure. After the film ruptured, the pressure was sharply decreased due to the higher-pressure difference. Meanwhile, the (dP/dt) was decreased with the increase of venting distance as the explosion venting progressed. Two peaks were decreased from 21.9 MPa s^{-1} and −61.1 MPa s^{-1} to 11.6 MPa s^{-1} and −11.7 MPa·s^{-1}, respectively. This is because that the energies of shock wave and venting flame were rapidly decreased under the combined action of heat dissipation and resistance. Figure 7.13 shows the effect of film rupture pressure on (dP/dt) histories at the venting port. The $(dP/dt)_{max}$ was evidently increased as the film rupture pressure enhanced, but the corresponding moment was prolonged. And an obvious negative value was also observed at the venting port and the drop extent continued to increase. This is because that the increase of film rupture pressure resulted in the greater venting pressure difference, further leading to the higher venting rate. Especially, the change of (dP/dt) history at the venting port was more significantly. Besides, the pressure history appeared a slight increase phenomenon in the later stage. This is because that the continuous increase of temperature inside the vessel further caused the high temperature gas to be released again through the venting duct. Combined with Table 7.3, the (dP/dt) history showed the same change trend as the rupture pressure increased.

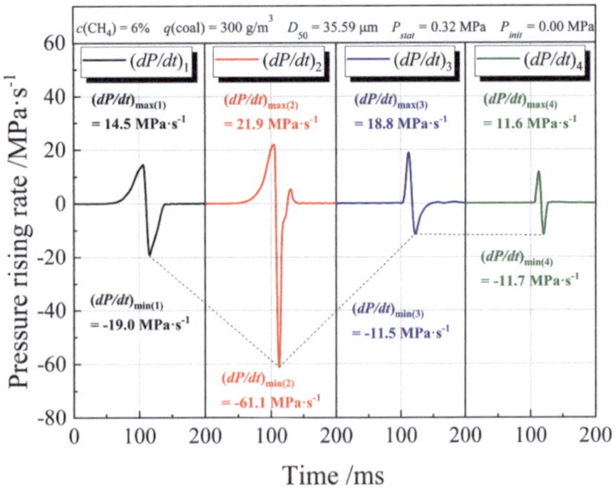

Fig. 7.12 Comparisons of (dP/dt) histories at different positions ($P_{stat} = 0.32$ MPa)

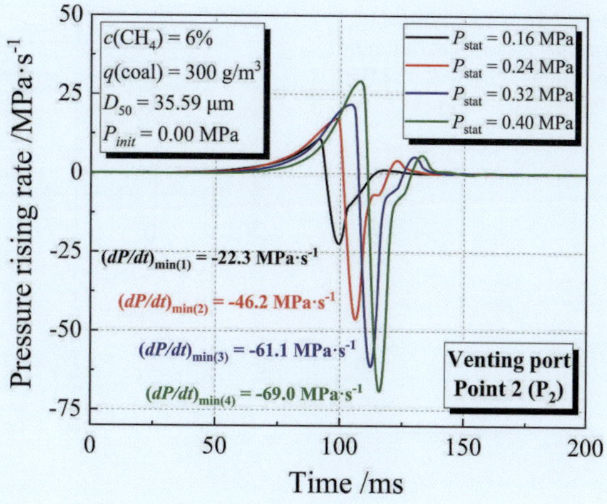

Fig. 7.13 Comparisons of (dP/dt) histories at venting port under different film rupture pressures

Table 7.3 Effect of film rupture pressure on (dP/dt) and their appearance moments

Film rupture pressure/MPa	Experimental data	Monitoring position			
		Point 1	Point 2	Point 3	Point 4
$P_{stat} = 0.16$	$(dP/dt)_{max}$/(MPa s^{-1})	7.68	10.89	9.85	2.72
	Appearance moment/ms	92.84	91.54	99.75	102.53
	$(dP/dt)_{min}$/(MPa s^{-1})	−7.70	−22.29	−5.44	−4.04
	Appearance moment/ms	104.65	99.39	110.46	110.12
$P_{stat} = 0.24$	$(dP/dt)_{max}$/(MPa s^{-1})	12.53	17.39	16.28	7.96
	Appearance moment/ms	99.48	98.31	109.89	111.25
	$(dP/dt)_{min}$/(MPa s^{-1})	−14.36	−46.20	−9.17	−7.99
	Appearance moment/ms	109.92	106.17	121.73	118.45
$P_{stat} = 0.32$	$(dP/dt)_{max}$/(MPa s^{-1})	14.54	21.88	18.78	11.60
	Appearance moment/ms	105.74	104.25	112.06	112.51
	$(dP/dt)_{min}$/(MPa s^{-1})	−19.04	−61.13	−11.52	−11.70
	Appearance moment/ms	115.83	112.27	121.92	119.93
$P_{stat} = 0.40$	$(dP/dt)_{max}$/(MPa s^{-1})	25.17	29.32	29.78	16.66
	Appearance moment/ms	109.25	107.9	116.53	116.82
	$(dP/dt)_{min}$/(MPa s^{-1})	−27.29	−68.98	−18.57	−7.56
	Appearance moment/ms	119.22	115.87	128.26	131.35

Effect of Initial Pressure

Figure 7.14 is the comparisons of (dP/dt) histories at different positions under 0.02 MPa initial pressure ($P_{stat} = 0.32$ MPa). Similar to the atmospheric condition, (dP/dt) history showed a change pattern of rising firstly and then reducing from

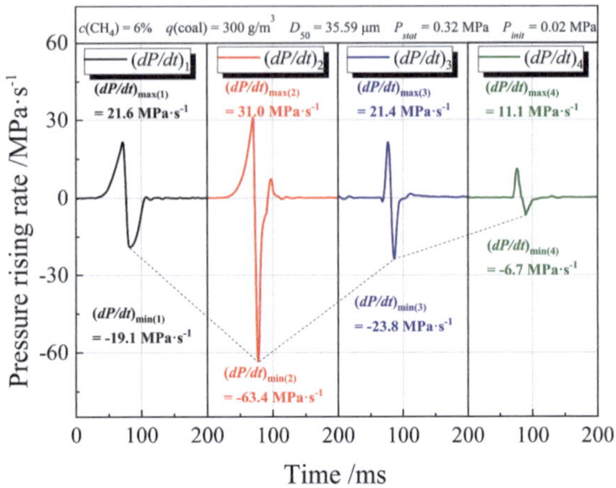

Fig. 7.14 Comparisons of (dP/dt) histories at different positions ($P_{stat} = 0.32$ MPa; $P_{init} = 0.02$ MPa)

the vessel interior to exterior as the venting progressed. The change of (dP/dt) was most significant at the venting port and it was reduced with the increase of venting distance. Especially for the changes in the two peaks.

Figure 7.15 shows the effect of initial pressure on (dP/dt) histories at the venting port ($P_{stat} = 0.32$ MPa). As the initial pressure increased, the moment for (dP/dt) history rise was advanced and its slope was increased. $(dP/dt)_{max}$ and $(dP/dt)_{min}$ were evidently increased and their corresponding moment were advanced. Combined with Table 7.4, the same change trend was also presented at different venting positions with the increase of initial pressure and the change at the venting port was most significant. This indicates that the rise of initial pressure led to an evidently reduction of molecular distance and increase of collision probability, resulting in the enhancement of explosion and increase of venting pressure difference under the determined film rupture pressure. The macroscopically presented the more significant pressure change.

7.2.4 Venting Flame Temperature

Temperature Field Distribution Characteristics

The distribution change of temperature field could directly reflect the explosion venting characteristics and flame disaster-causing mechanism. Figure 7.16 presents the correspondence of temperature distribution and venting flame image during the venting at atmospheric pressure ($P_{stat} = 0.40$ MPa). Both of them exhibited the same morphology and change trend. The flame was the brightest and presented a clear outline at 0.25–6.25 ms, and the flame shape and high temperature distribution image were completely similar at this stage. This indicates that the flame was accelerated to release under high pressure difference and accompanied by a reaction of unburned

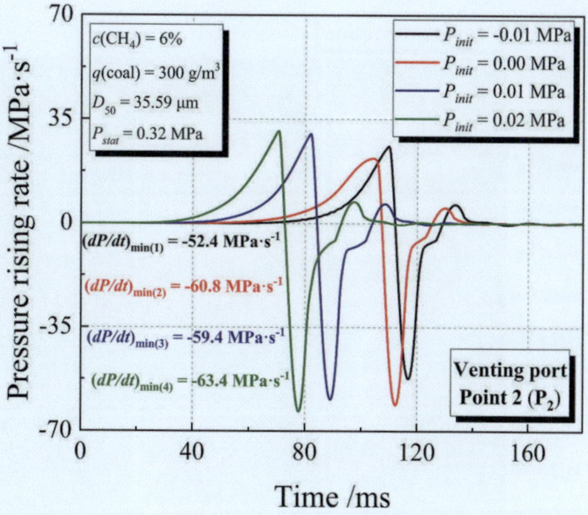

Fig. 7.15 Comparisons of (dP/dt) histories at venting port under different initial pressures ($P_{stat} = 0.32$ MPa)

Table 7.4 Effect of initial pressure on (dP/dt) and their appearance moments

Initial pressure/MPa	Experimental data	Monitoring position			
		Point 1	Point 2	Point 3	Point 4
$P_{init} = -0.01$	$(dP/dt)_{max}/(MPa\ s^{-1})$	18.65	25.93	13.63	10.57
	Appearance moment/ms	110.22	109.75	116.81	115.44
	$(dP/dt)_{min}/(MPa\ s^{-1})$	−16.39	−52.35	−13.35	−5.81
	Appearance moment/ms	120.06	116.93	122.90	124.27
$P_{init} = 0.00$	$(dP/dt)_{max}/(MPa\ s^{-1})$	14.54	21.88	18.78	11.60
	Appearance moment/ms	105.74	104.25	112.06	112.51
	$(dP/dt)_{min}/(MPa\ s^{-1})$	−19.04	−61.13	−11.52	−11.70
	Appearance moment/ms	115.83	112.27	121.92	119.93
$P_{init} = 0.01$	$(dP/dt)_{max}/(MPa\ s^{-1})$	21.17	30.08	15.15	11.36
	Appearance moment/ms	82.37	81.86	87.83	87.80
	$(dP/dt)_{min}/(MPa\ s^{-1})$	−18.63	−59.36	−32.62	−5.75
	Appearance moment/ms	91.81	88.95	94.09	100.79
$P_{init} = 0.02$	$(dP/dt)_{max}/(MPa\ s^{-1})$	21.55	30.96	21.35	11.08
	Appearance moment/ms	71.27	70.39	76.71	76.30
	$(dP/dt)_{min}/(MPa\ s^{-1})$	−19.06	−63.35	−23.75	−6.71
	Appearance moment/ms	82.13	77.73	85.99	89.90

substances (the acceleration venting stage). Under the actions of drag and buoyancy, the high temperature flame appeared an upward tilt process (t = 4.25 ms). At 6.25 ms, the venting flame length and the temperature distribution area reached the largest. Afterwards, the flame was gradually extinguished, and its length was constantly reduced because of the heat dissipation. Nevertheless, the extinguished flame still

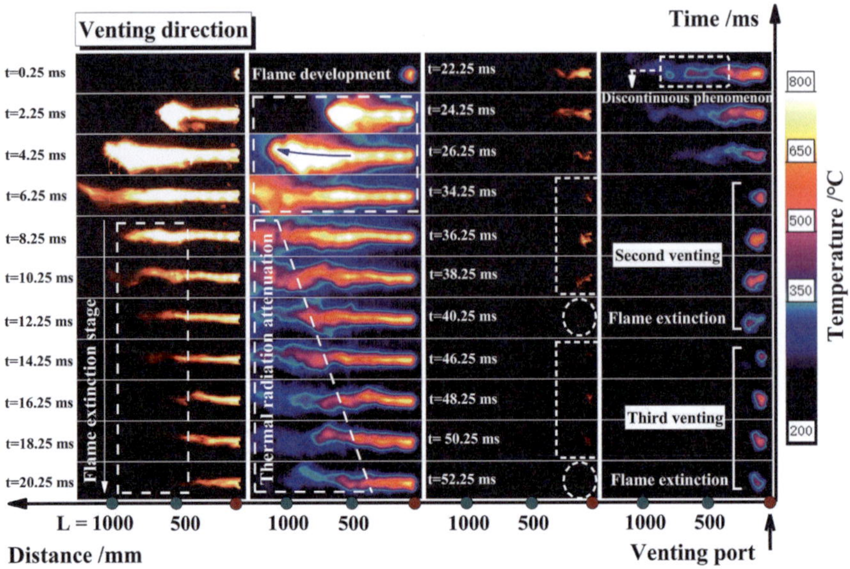

Fig. 7.16 Changes of venting flame propagation and temperature field distribution during the venting at atmospheric pressure ($P_{stat} = 0.40$ MPa)

maintained the higher temperature, resulting in a larger distribution outline than the flame shape. But the temperature of flame front area was evidently reduced. Due to the continuous venting of explosion products, the high temperature flame near the venting port remained bright and gradually decreased. At 22.25 ms, the flame presented evidently the oscillation and formed a discontinuous structure as a result of the thermal dynamic instability. Besides, the flame emerged the second and third venting phenomenon in the later stage. This is because that the temperature of vessel interior continued to increase after extinguishing, causing the once more venting of flame and exhibiting a weak cluster flame structure under the pressure difference. Hence, the venting port appeared a visible high temperature area (range: 350–500 °C). Thereafter, the flame was thoroughly extinguished, and the high temperature area was disappeared.

Effect of Film Rupture Pressure

Figure 7.17 illustrates the effect of film rupture pressure on the flame temperature distribution under the atmospheric pressure. As $P_{stat} = 0.16$ MPa, the venting flame exhibited a "strip" structure as the venting developed and the temperature near the venting port was higher under the lower pressure difference. Due to the heat dissipation, the temperature of venting flame away from venting port was obviously reduced. As the rupture pressure enhanced, the unburned mixture was discharged to vessel exterior and reacted, leading to an increase in the temperature of flame front area.

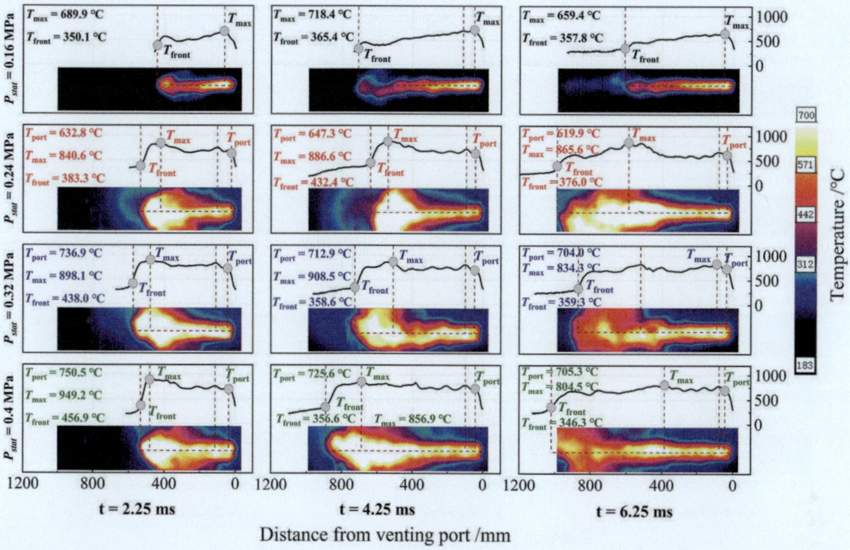

Fig. 7.17 Effect of film rupture pressure on the temperature field distribution at atmospheric pressure

Under high pressure difference, the flame front area became the larger and presented an upward trend. T_{front}, $T_{\max}$ and T_{port} were also sequentially increased (t = 2.25 ms). $T_{\max}$ appeared in the center of "spherical" flame near front and their distance was increased as the venting progressed. Thereafter, the heat was rapidly dissipated. And the flame temperature corresponding to the maximum film rupture pressure appeared firstly a decreasing trend. At 4.25 ms, T_{front} under 0.32 and 0.40 MPa film rupture pressures was approximate, but it was increased with the decrease of film rupture pressure. And $T_{\max}$ was also increased as the film rupture pressure decreased at 6.25 ms.

Effect of Initial Pressure

Figure 7.18 presented the effect of initial pressure on the flame temperature distribution (P_{stat} = 0.32 MPa). At 2.25 ms, the flame was acceleratingly released. As the initial pressure increased, T_{front}, $T_{\max}$ and T_{port} all presented an increasing trend. And $T_{\max}$ did not appear on the flame front, but appeared in the central area near the cluster flame. As the flame continued to venting (t = 4.25 ms), the temperature did not appear an evident change at the venting port as $P_{\text{init}} < 0.02$ MPa. As $P_{\text{init}} = 0.02$ MPa, the temperature at the venting port presented a slight increase trend. However, T_{front} and $T_{\max}$ were all evidently increased as the initial pressure enhanced. This indicates that the flame was still in an acceleration venting stage at this moment, and the temperature continued to increase as the initial pressure

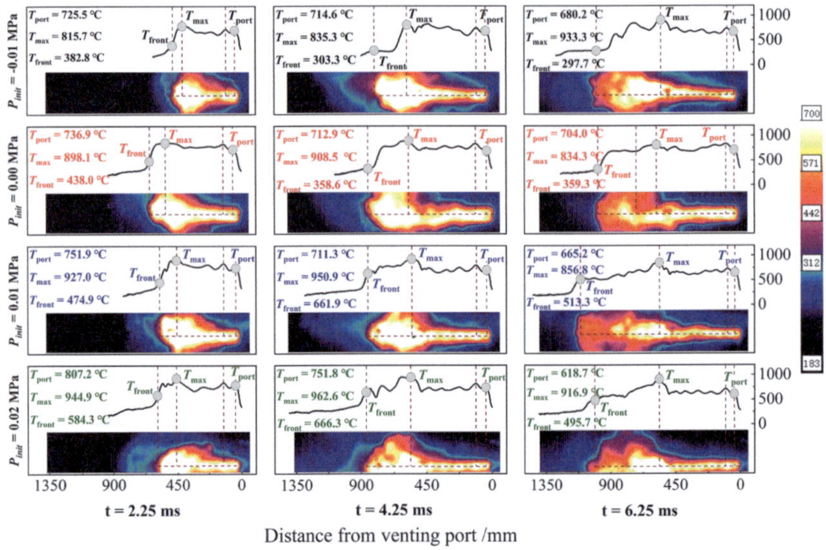

Fig. 7.18 Effect of initial pressure on the temperature field distribution (P_{stat} = 0.32 MPa)

enhanced. With progress of the venting (t = 6.25 ms), the venting flame with higher initial pressure began to decay and T_{front} was decreased. However, T_{max} was still increased as the initial pressure enhanced. As $P_{init} = -0.01$ MPa, the duration of flame acceleration venting was prolonged due to the relatively small venting pressure difference, resulting in a relatively high venting flame temperature at 6.25 ms (T_{max} = 933.3 °C). The above indicates that the temperature distribution of venting flame was related to the mixture explosion intensity inside the vessel, and affected by the initial pressure. The enhancement of initial pressure could increase the temperature of venting flame, but the time for the flame began to extinguish was advanced.

7.3 Conclusions

By analyzing the relationships between the venting parameters of both the vessel interior and exterior, the dynamic venting characteristics of gas/dust two-phase mixture explosions were studied. The effect regulars of film rupture and initial vessel pressures on the explosion venting parameters were studied, and the influence mechanism of operation condition on the dynamic characteristics of mixture explosions venting were revealed.

First, the correspondence between the pressure rise and flame propagation of both the vessel interior and exterior during mixture explosion venting was described. The venting flame experienced a transformation process of "mushroom", "spherical" and "strip" structures and was affected by the operation condition. As P_{stat} = 0.16 MPa, the flame always propagated in a small "strip" structure and its venting length was the shortest. As $P_{stat} \geq 0.24$ MPa, the flame all exhibited a "mushroom" structure and it became more prominent as the film rupture pressure enhanced. The velocity history was also evidently increased in the early stage as the film rupture and initial pressures enhanced, and the flame exhibited an obvious oscillation during the extinguishing stage and the average amplitude was also enlarged. The flame appeared a secondary venting phenomenon as $P_{stat} \geq 0.32$ MPa. Especially, the third venting phenomenon appeared as $P_{stat} = 0.40$ MPa.

Second, the venting pressure and its rise rate from the interior to exterior of vessel presented a change trend of rising firstly and then reducing during the venting. And the changes in the pressure rise/drop at the venting port got more significant due to the sudden expansion of venting space and acceleration of duct. With increases of initial and film rupture pressures, the explosion venting intensity were obviously increased. And a larger negative pressure was also observed at the venting port and its reduction extent continued to increase. Simultaneously, the moment when the negative pressure appears corresponded to the flame oscillation.

Third, the changes of venting flame image and temperature distribution in the acceleration venting stage exhibited the same morphology and change trend. Afterwards, the flame was gradually extinguished and its length was constantly reduced because of the heat dissipation. Nevertheless, the extinguished flame still maintained a higher temperature, resulting in a larger distribution outline than the flame image. But the temperature in the flame front area was evidently decreased and became the discontinuous structure. T_{max} appeared in the center of "spherical" flame near front and their distance was increased as the venting progressed. The increases of film rupture and initial pressures could enhance the temperature of venting flame, but the corresponding flame temperature appeared firstly the decreasing trend. Besides, the flame temperatures of the second and third venting could reach the range of 350–500 °C.

To achieve effective protection for the explosion venting of two-phase systems, three suggestions are proposed based on the experiment results: (1) The corresponding formula of venting parameter under different the operation conditions with the changes of temporal and spatial distribution should be established. (2) The design of venting device should fully consider the overpressure generated by the acceleration effect of venting duct. (3) The flame temperatures of secondary and multiple venting could reach the ignition temperature of highly active gas, and the occurrence of secondary venting should be avoided as much as possible by changing operation condition.

References

1. Cao XY, Huang R, Lu YW, Wang ZR, Fan LT, Zhou YQ, Xu JJ. Experimental research on the venting characteristics of methane and coal dust two-phase system mixture explosions. Fuel 2023; 353: 129208.

2. Rui S, Guo J, Li G, Wang C. The effect of vent burst pressure on a vented hydrogen-air deflagration in a 1 m^3 vessel, Int. J. Hydrog. Energy 2018; 43: 21169–21176.
3. Wang S, Zhao Y, Li G, Xie Y, Wu D. Effect of different vent covers on the overpressure and flame propagation characteristics of hydrocarbon fuel-air mixture venting explosion, Fuel 2022; 324: 124620.
4. Schiavetti M, Carcassi M. Analysis of acoustic pressure oscillation during vented deflagration and proposed model for the interaction with the flame front, Int. J. Hydrog. Energy 2016; 42: 7707–7715.
5. Sun S, Wang M, Guo K, Zhao T, Guo Q. Effect of vent conditions on internal overpressure time-history during a vented explosion, J. Loss Prev. Process Ind. 2018; 54: 85–92.

Chapter 8
Mechanisms of Vapor Cloud Explosion and Its Chain Reaction Induced by an Explosion Venting Flame

8.1 Experimental Apparatus and Procedures

Figure 8.1 shows the schematic diagram of the experimental apparatus used to analyze the vapor cloud explosion and its chain reaction induced by the venting flame. The apparatus includes a venting container, a vapor cloud explosion device, a gas supply system, an ignition system, an image acquisition system, a pressure acquisition system, a temperature acquisition system, and a procedure control and data acquisition system.

The venting container (422 mm × 422 mm) had a good explosion resistance (Q345R plate). The pressure sensor, the inlet and outlet valves, and the ignition rod were installed at the center of the container sidewall and distributed in intervals of 90° along the circumferential direction. A duct ($L = 125$ mm) was installed at the right end of the container for the vent of the explosion flame. And the duct was open before each ignition. The vapor cloud explosion device consists of a 0.05 mm-thick polyvinyl chloride hemispherical film and an aluminum alloy bottom plate. A pressure sensor and a premixed gas inlet valve were installed at the bottom plate, as shown in Fig. 8.2. The gas supply system consists of a vacuum pump (2X-8GA), an automatic valve distribution instrument (RCS2000-B), a methane cylinder, and an air compressor (DA-7). An igniter with 5–20 J output energy was adopted to ignite the premixed gas. A high-speed camera and a schlieren system (ϕ 300 mm) were used to capture the flame propagation process and the flame structure, respectively. A pressure sensor with a dynamic responding time of 1 ms (0–5 MPa; +0.25% FS) and three pressure transmitters (0–0.7 MPa; 300 kHz) were used to acquire the pressure data inside and outside the container, respectively. A high-speed infrared thermal imager was adopted to collect the flame temperature data (0–2500 °C; ±1 °C). The minimum detection difference and the maximum frame frequency were 18 mK and 5600 Hz, respectively. A thermocouple with a platinum–rhodium alloy wire was also used to record the flame temperature data (±1.5 °C; ±0.25% t). A high-frequency data acquisition card (SIRIUS), with high speed parallel analog inputs and programmable

Z. Wang and X. Cao, *Mechanism of Composite Explosion Inhibition and Venting Protections*, https://doi.org/10.1007/978-981-95-4809-5_8

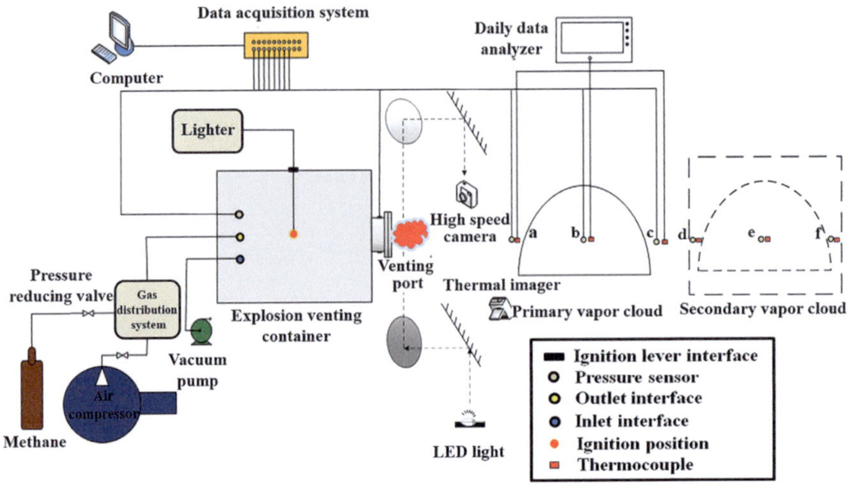

Fig. 8.1 Schematic diagram of the experimental apparatus

digital outputs, was adopted to realize the ignition and the data acquisition in proper sequence.

The premixed gas of a certain methane concentration was selected to carry out the experiments of the vapor cloud explosion and its chain reaction induced by the venting flame. The premixed gas was ignited with 5 J energy at the center of the container. The initial pressure was 0.1 MPa. Two vapor clouds were installed

Fig. 8.2 Vapor cloud explosion device

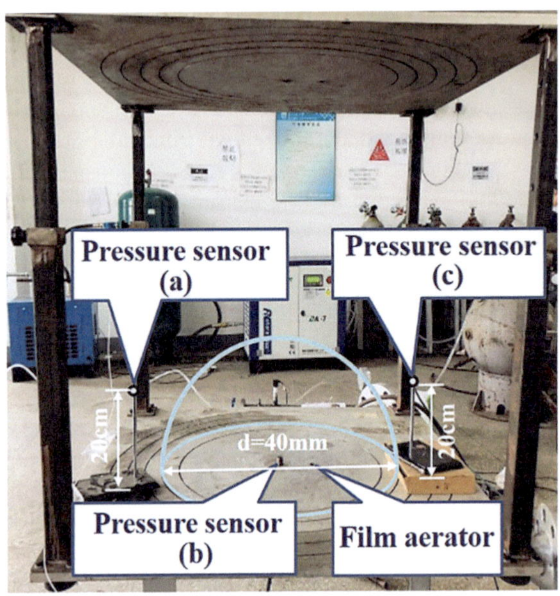

Table 8.1 Explosion venting experimental scheme

Container volume (L)	Venting port diameter	Venting area ratio	Methane concentration
55	59	100%	11%
Vapor cloud methane concentration	Primary and secondary vapor cloud diameter (mm)	Distance between the primary vapor cloud center point and venting port (mm)	Distance between the center points of two vapor clouds (mm)
11%	400	400	100

successively in front of the explosion venting container. These two vapor clouds were named primary vapor cloud (the vapor cloud ignited by the venting flame) and secondary vapor cloud (the vapor cloud ignited by the primary vapor cloud). Pressure and temperature sensors were installed at the boundaries and centers of the primary (points a, b, and c) and secondary (points d, e, and f) vapor clouds, respectively. The specifics of the experimental scheme are listed in Table 8.1.

8.2 Results and Discussions

8.2.1 Vapor Cloud Explosion Induced by the Explosion Venting Flame

Flame Structural Characteristics

The development process of the flame structure with time during the vapor cloud explosion induced by the venting flame is shown in Fig. 8.3. The moment shown in the figure corresponds to the relative time in which the flow field is affected by the pressure wave. A clear boundary between the vapor cloud film and the external flow field was observed in the early stage. At 2 ms, the high-pressure unburned gas appeared in the observation region, and it first came in contact with the vapor cloud film and caused film deformation inward [1]. Subsequently, a venting flame emerged at 3 ms and came in full contact with the vapor cloud film at 4 ms. At this moment, the vapor cloud film remained intact, and the vapor cloud was not ignited. After 4 ms, the vapor cloud film ruptured (t = 8 ms), resulting in the outflow of the premixed gas (t = 10 ms). Then, the vapor cloud was ignited and the flame expanded rapidly. Although the flame presented an irregular structure and became blurry due to the change in flow field density, an interface between the venting flame and the vapor cloud film was clearly presented. Subsequently, the flame propagated forward under the action of the venting pressure, and the vapor cloud volume was gradually reduced until the flame disappeared (t = 10–15 ms). At 15 ms, the flame structure became clear. As the flame propagated ceaselessly, the irregular flame structure near

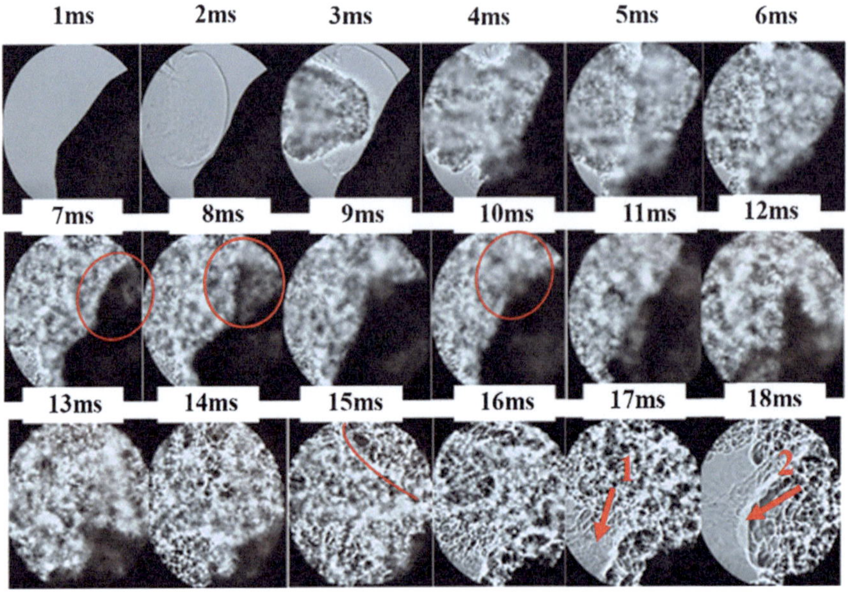

Fig. 8.3 Flame structure evolution process of vapor cloud explosion induced by the venting flame

the left side (point 1) gradually became smooth and appeared a clear boundary. The boundary moved forward with flame propagation (point 2), indicating that the vapor cloud explosion flame and venting flame as a whole propagated at this moment and tend to be a steady state under the action of the venting pressure.

Figure 8.4 presents the propagation process of the vapor cloud explosion flame induced by the venting flame. The moment shown in the figure corresponds to the relative time in which the vapor cloud film is ruptured. The "blue" venting flame propagated steadily forward and came in contact with the vapor cloud film, and it propagated further in the upper region after entering the vapor cloud film (t = 3–5 ms). This phenomenon described one of the propagation characteristics of the venting flame. The propagation characteristics could also be attributed to the density of the methane, which was less than that of air inside the vapor cloud film. Then, the area above the oblique side of the vapor cloud film first ruptured due to the thermal action of the flame, and flame was released from the ruptured area (point 1). As the flame developed, the ruptured area widened continuously (points 2 and 3). The flame brightness was enhanced and changed from pale-blue to yellow-blue, indicating that vapor cloud was ignited. As the combustion reaction proceeded further, the explosion flame became more intense and expanded gradually (t = 11–12 ms). After that, the flame changed into yellow-white, and the combustion region was reduced (t = 12–18 ms). It indicates that the premixed gas in this region achieved complete combustion under the action of the two flames, and then the explosion flame was extinguished gradually [2].

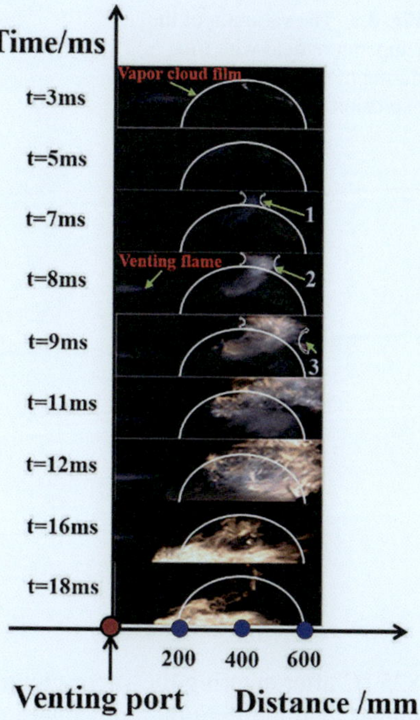

Fig. 8.4 Flame propagation process of vapor cloud explosion induced by the venting flame

Flame Temperature Characteristics

Variation in the temperature field with time can directly reflect the mechanism of the vapor cloud explosion induced by the venting flame [3]. Figure 8.5 shows the variation process of the temperature field with time, as recorded by the high-speed thermal imager. The moment in the figure corresponds to the relative time in which the flame is captured. At 1 ms, the temperature of the vapor cloud film was increased (point 1) due to the action of flame thermal radiation. Subsequently, the venting flame entered the vapor cloud film (t = 7 ms) and ignited the vapor cloud (t = 9 ms). However, the temperature of the vapor cloud explosion flame was relatively low at this moment, and the vapor cloud was not ignited instantly. As the flame propagated, the volume of the vapor cloud explosion flame expanded ceaselessly. The flame presented an upward trend due to the actions of thermal buoyancy and venting pressure (t = 11–13 ms). And the brightest region was observed at the center of the vapor cloud (point 2). The reaction reached the highest intensity at 15 ms, corresponding to the complete combustion of the vapor cloud. This phenomenon also indicates that the vapor cloud could be completely ignited by the venting flame after entering the vapor cloud film, at which it further developed for a certain period. And the highest temperature appeared in the central region of the vapor cloud.

Figure 8.6 presents the corresponding relationship between the temperature field and temperature curve along the horizontal direction of the venting port. The maximum temperature (T_{max}) of the "mushroom cloud" flame center was as high as

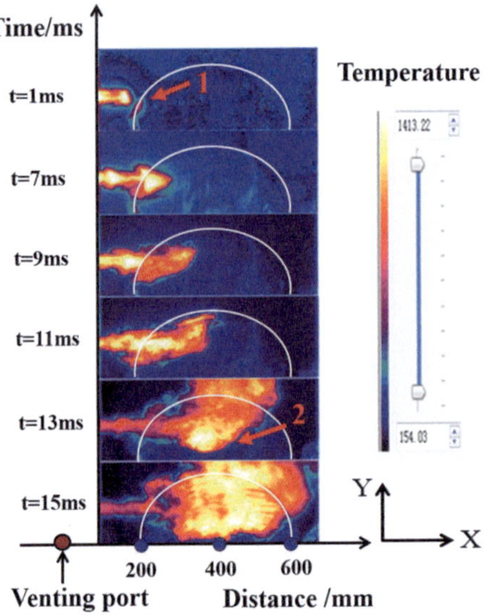

Fig. 8.5 The variation of the temperature field with time during the vapor cloud explosion

825 °C after entering the vapor cloud film (Fig. 8.6a). This temperature was much higher than the ignition temperature of the methane (538 °C). However, the flame retained its original shape and temperature, indicating that the vapor cloud had not been ignited. It also indicates that temperature was not the only factor to induce vapor cloud explosion. As for the ignition condition, the venting flame, which came in contact with the vapor cloud, needed to provide the required energy for the explosion. After that, although the flame shape changed slightly and the flame temperature decreased (T_{max} = 775 °C), the vapor cloud presented a burned trend (Fig. 8.6b). Subsequently, the flame temperature inside the vapor cloud was increased, and the highest temperature reached 925 °C (Fig. 8.6c), indicating that the vapor cloud was locally ignited. As the flame developed ceaselessly, the temperature was increased further (T_{max} = 1050 °C) (Fig. 8.6d). The main reason for this change was that the vapor cloud was integrally ignited and the flame propagated outward as a spherical structure.

Figure 8.7 shows the variation of temperature curves with time at different locations of the primary vapor cloud (points a, b, and c) during the vapor cloud explosion induced by the venting flame. The moments of temperature began to rise at 270 ms, 590 ms, and 865 ms with the increase in the distance from the venting port. However, the T_{max} of the three points (a, b, and c) presented a trend of rising firstly and then decreasing. The highest temperature was observed in the central region of the vapor cloud (T_{max} = 1450 °C). This temperature was far higher than those in point a (T_{max} = 920 °C) and point c (T_{max} = 901 °C), indicating that the combustion intensity in the central region of the vapor cloud was enhanced under the actions of the venting flame and the vapor cloud explosion flame. Meanwhile, the temperature curve appeared an oscillation process at peak value due to the interaction of the both flames. As the

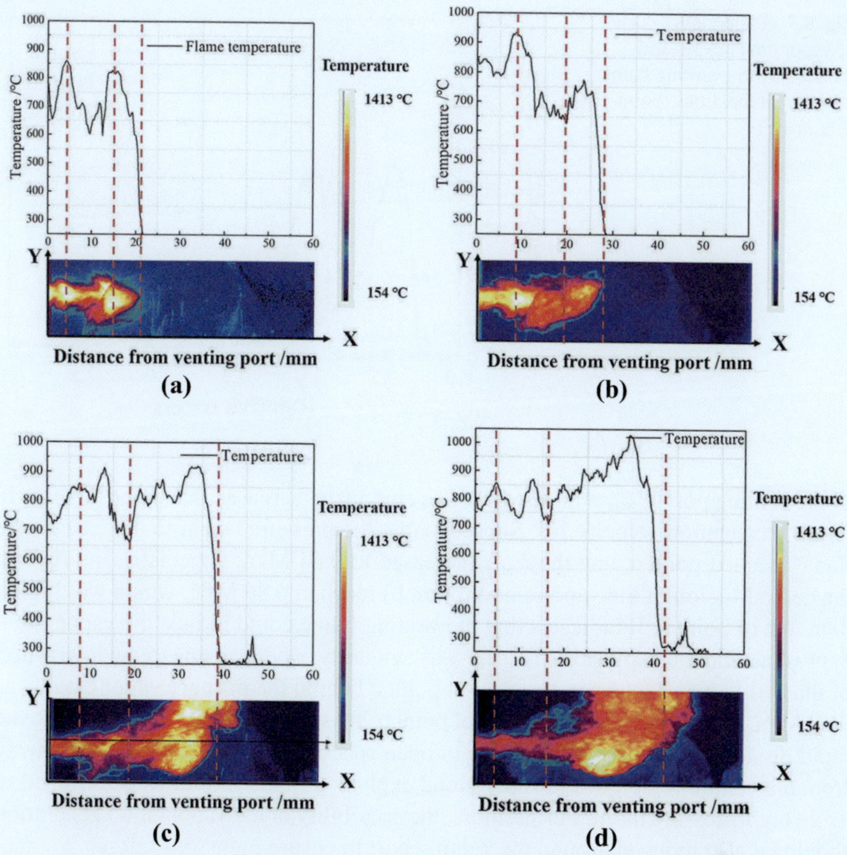

Fig. 8.6 The corresponding relationship between the temperature field and temperature curve of vapor cloud explosion (**a** 7 ms; **b** 9 ms; **c** 12 ms; **d** 13 ms)

flame propagated, the temperature decreased significantly due to the reduction in explosion intensity (point c). Hence, it should be as far away from the central region of the vapor cloud as possible to avoid high-temperature harm after the vapor cloud was ignited by the venting flame.

Pressure Distribution Characteristics

Figure 8.8 presents the variation of pressure curves at different locations (inside the container, the venting port, and points a, b, and c) during the vapor cloud explosion induced by the venting flame. The explosion pressure increased firstly and then decreased with the increase in the distance from the venting port. The maximum explosion pressure (P_{max}) was observed at point b. After ignition, although the pressure inside the container was increased ($P_{max} = 0.35$ MPa), the rising rate was relatively small. The flame and pressure wave could not be expelled rapidly due to the confined propagation space, resulting in pressure curve oscillation [1]. The pressure

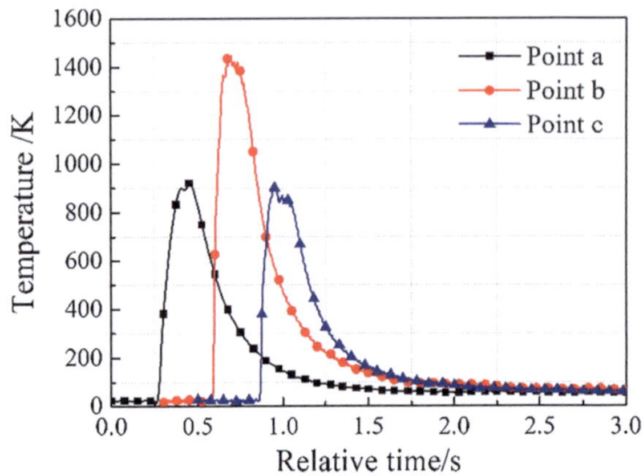

Fig. 8.7 Temperature curve of vapor cloud explosion induced by the venting flame at different locations (point a, b, and c)

of the venting port ($P_{max} = 0.43$ MPa) was obviously increased due to the increase in flame propagation velocity [4]. Subsequently, the pressure began to increase as the flame reached point a, and the P_{max} increased to 0.64 MPa. Especially, the P_{max} in the central region of the vapor cloud (point b) reached 0.86 MPa, which was higher than that of point a. It indicates that the venting flame could induce the vapor cloud explosion, and the explosion intensity was evidently enhanced due to the interaction of the two flames. However, the P_{max} at point c behind the primary vapor cloud was 0.62 MPa, which was lower than that of point b. This reason could be attributed to the rapid dissipation of explosion energy in open space. Hence, the safe distance away from the central region of the vapor cloud explosion flame should be determined to avoid high-pressure harm. Furthermore, the possibility of the vapor cloud generation should be also reduced around the venting port to ensure industrial safety.

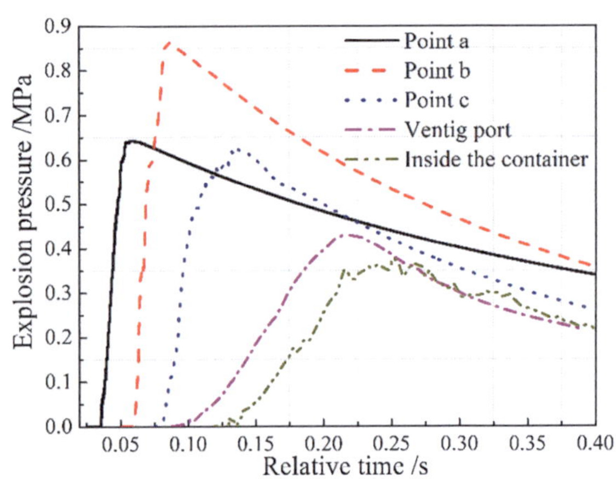

Fig. 8.8 Pressure curve of vapor cloud explosion induced by the venting flame at different locations

8.2.2 Vapor Cloud Explosion Chain Reaction Induced by the Explosion Venting Flame

Flame Structure Characteristics

Figure 8.9 presents the change of the flame structure with time during the vapor cloud explosion chain reaction induced by the venting flame. The capture region was located between the primary and secondary vapor clouds. The moment in the figure corresponds to the relative time in which the primary vapor cloud film occur deformation. It shows that the vapor clouds film were intact at 1 ms, and a clear boundary between the flow field and the vapor cloud film was presented. At 2 ms, the volume of the primary vapor cloud began to expand due to the action of the explosion flame. Nevertheless, its film was still intact and the secondary vapor cloud film was also unaffected. The flame propagated forward with a "fan-shaped" structure after the primary vapor cloud film ruptured at 3 ms and came in contact with the secondary vapor cloud film at 5 ms. Under the action of high temperature flame, the secondary vapor cloud film did not ruptured instantly (t = 6–9 ms). The film began to expand at 10 ms and ruptured at 11 ms. After that, the internal premixed gas was ignited and appeared with a blurry flame structure (t = 11–13 ms). After 14 ms, the explosion flames of the two vapor clouds mixed evenly with the continuous development of explosion flame. Accompanied by the gradual disappearance of the irregular flame structure, the flame structure became uniform and dense (t = 15–18 ms). The change process of flame structure indicates that the venting flame could induce vapor cloud explosion and set off a chain reaction. As for the induction condition, the vapor cloud explosion flame, which came in contact with the secondary vapor cloud, needed to provide the required energy for the explosion.

Flame Temperature Characteristics

Figure 8.10 shows the variation process of the temperature field with time during the vapor cloud explosion chain reaction induced by the venting flame. The moment in the figure corresponds to the relative time in which the central region of the primary vapor cloud is ignited. As shown in Fig. 8.10, the temperature in central region of the primary vapor cloud was as high as 925 °C, indicating a localized ignition by the venting flame. As the explosion proceeded, the high temperature area widened gradually. The flame propagated forward and reached the secondary vapor cloud under the action of the venting pressure (t = 6–8 ms). Due to the reduction of venting pressure, the temperature field near the secondary vapor cloud presented an attenuation trend (t = 12 ms), indicating that the secondary vapor cloud was not ignited at this moment. After that, a bright region appeared in front of the explosion flow field and widened gradually (t = 14–15 ms). This phenomenon indicates that the secondary vapor cloud was ignited, and an explosion chain reaction occurred. Furthermore, an interface was observed between the two flames (points 1 and 2 in

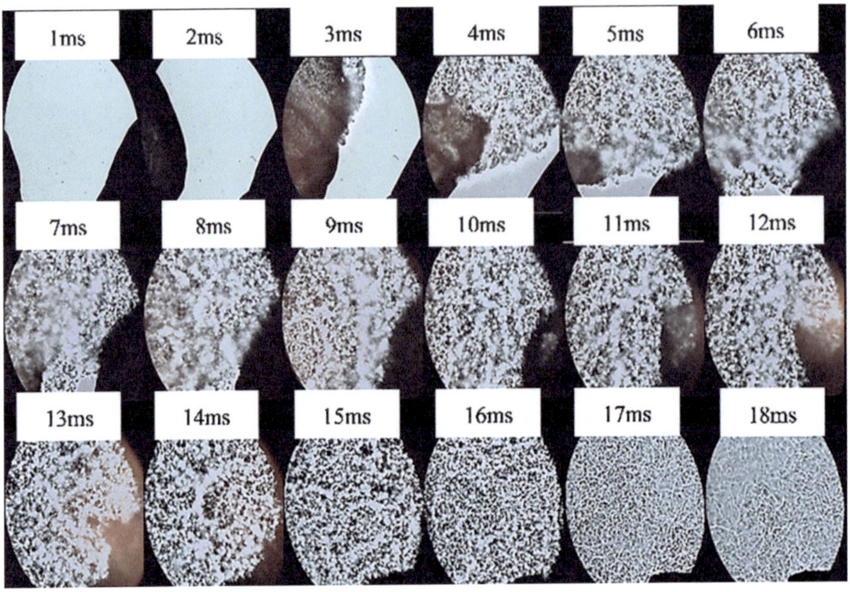

Fig. 8.9 Flame structure evolution process of vapor cloud explosion chain reaction induced by the venting flame

Fig. 8.10). As the flame propagated, the two flames developed into a bright flame, and the explosion reaction was enhanced gradually (t = 15–17 ms). Although the venting process of the container was still ongoing, the temperature of the venting flame was lower than that of the vapor cloud explosion flame.

Figure 8.11 shows the corresponding relationship between the temperature field and temperature curve of the vapor cloud explosion chain reaction along the horizontal direction of the venting port. It can be seen that the temperature of the flame center region was as high as 1300 °C in Fig. 8.11a. Although the temperature near the secondary vapor cloud increased to 927 °C, the temperature of this region presented a decrease trend (from 8 to 12 ms in Fig. 8.10). It indicates that, as a result of instantaneous high temperature of the explosion flame, the secondary vapor cloud film did not rupture. Subsequently, the flame temperature was increased and the highest temperature reached 1475 °C. The temperature near the secondary vapor cloud was also increased and reached 1226 °C (Fig. 8.11b). The results indicate that the secondary vapor cloud film occurred rupture and the vapor cloud was ignited due to the continuous action of flame thermal radiation. Especially, the temperature near the secondary vapor cloud exceeded 1411 °C in Fig. 8.11c, indicating that the secondary vapor cloud was completely ignited and reacted violently. It also means that the chain reaction of the vapor cloud explosion could set off a high-intensity explosion. To avoid harm caused by the high temperature, the farthest distance away from the region where the chain reaction of the vapor cloud explosion could affect should be determined.

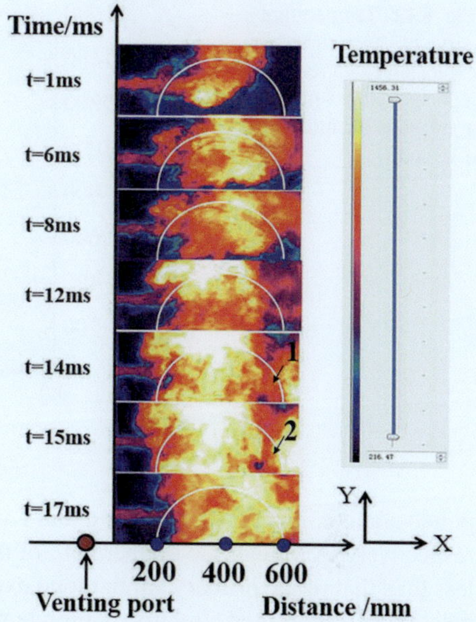

Fig. 8.10 The variation of the temperature field with time during the vapor cloud explosion chain reaction

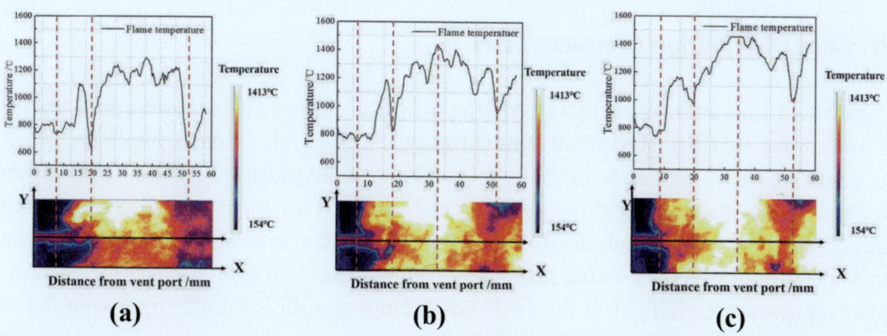

Fig. 8.11 The corresponding relationship between the temperature field and temperature curve of vapor cloud explosion chain reaction (**a** 12 ms; **b** 14 ms; **c** 15 ms)

Figure 8.12 shows the variation of temperature curves with time at different locations of the secondary vapor cloud (points d, e, and f) during the vapor cloud explosion chain reaction. Its variation trend was similar to that of the primary vapor cloud explosion (points a, b, and c). With the increase in the distance from the venting port, the moment in which temperature began to rise was prolonged from 75 to 95 ms and 142 ms. However, the temperature presented a trend of rising firstly and then decreasing. The highest temperature of the chain reaction was located in the central region of the secondary vapor cloud (T_{max} = 1398 °C), which was higher than that of point d (T_{max} = 901 °C) and point f (T_{max} = 767 °C). Besides, the T_{max} of point e was slightly lower (52 °C) compared with that of point b (Fig. 8.7). Compared with

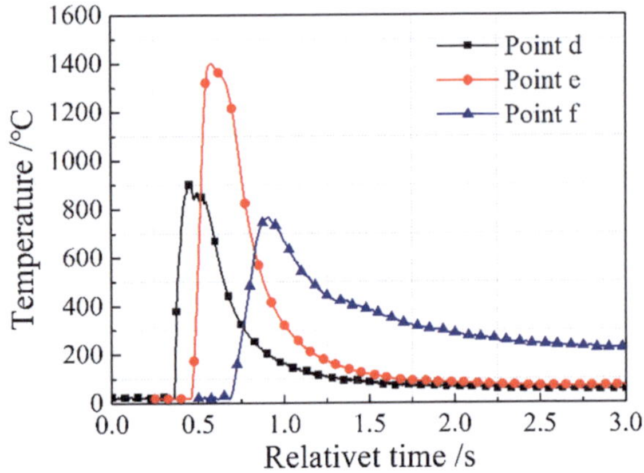

Fig. 8.12 Temperature curve of vapor cloud explosion chain reaction induced by the venting flame at different locations (point d, e, and f)

points a and c, the T_{max} of points d and f were also reduced by 24 °C and 134 °C, respectively. Although the temperature of the explosion chain reaction was slightly less than that of the primary vapor cloud explosion, the explosion flame of the chain reaction still presented a relatively high temperature.

Pressure Distribution Characteristics

Figure 8.11 illustrates the effect of hydrogen concentration on explosion pressure (H = 140 mm, $\varepsilon = 0.30$). Hydrogen concentration significantly impacts explosion pressure and its rising rate at both ends, influencing the explosion reaction rate, which in turn affects the propagation velocity and intensity of the shock wave entering the narrow channel. The relationship between hydrogen concentration and explosion pressures, as well as their rising rates at both ends, follows a similar variation pattern; however, pressure changes at the two ends are not significant under opposing flame resistance effects. At c = 12 and 15%, flame resistance is successful, with lower pressure exceeding upper pressure. Conversely, at c = 18 and 21%, the flame resistance fails, resulting in upper pressure being significantly higher than lower pressure. As hydrogen concentration increases, pressures at both ends exhibit a slight upward trend. This indicates that hydrogen concentration has a notable impact on the rates of pressure rise at both the front and rear ends of the flame-retardant system, with the likelihood of failure increasing as concentration rises.

Figure 8.13 shows the variation of pressure curves with time at different locations (inside the container, the venting port, and points d, e, and f) during the vapor cloud explosion chain reaction. The explosion pressure first increased and then decreased with the increase in the distance from the venting port. The highest explosion pressure was also observed in the central region of the secondary vapor cloud (point e; P_{max} = 0.89 MPa). However, the moment in which the pressure began to rise presented an opposite trend. After ignition, the pressure inside the container began to rise and appeared an oscillation process due to the action of the pressure wave. The explosion

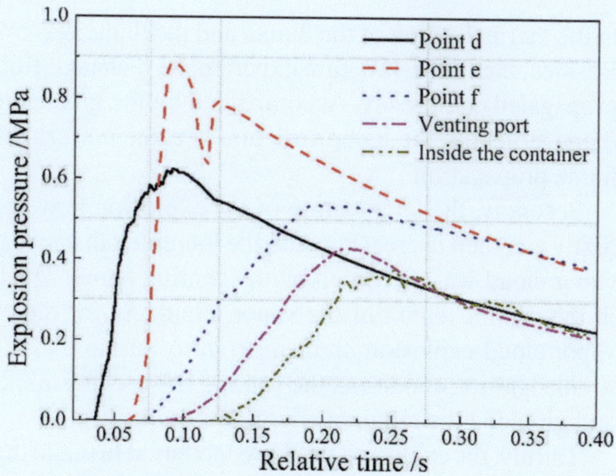

Fig. 8.13 Pressure curve of vapor cloud explosion chain reaction induced by the venting flame at different locations

flame acceleratingly propagated at the venting port under the action of the pressure gradient. The increase in flame propagation velocity was the main reason of the pressure rise [4]. Accompanied by the forward propagation of the primary vapor cloud explosion flame, the pressure of point d (P_{max} = 0.62 MPa) began to rise. Especially, the pressure in the central region (point e) was increased sharply after the secondary vapor cloud was ignited, and the P_{max} appeared. The pressure was increased again due to the energy provided continuously by the primary vapor cloud explosion flame, and the second pressure peak value reached 0.79 MPa. Meanwhile, the pressure at point f was also increased (P_{max} = 0.54 MPa) as the explosion flame propagated forward, but the rising rate was evidently decreased. The above data indicates that the vapor cloud explosion chain reaction could produce a high-intensity explosion and could not be ignored.

8.3 Conclusions

A set of experimental apparatus for vapor cloud explosion and its chain reaction induced by the venting flame was established in this study. It was found that the venting flame could induce a vapor cloud explosion and generate a chain reaction. By analyzing the changes in flame propagation characteristics, flame temperature, and pressure distribution, the mechanisms of vapor cloud explosion and its chain reaction induced by the venting flame were deeply investigated. Conclusions can be summarized as follows:

Firstly, the high-pressure unburned gas first came in contact with the vapor cloud film compared with the venting flame and then caused an inward film deformation. The film did not rupture instantly after contacting with the venting flame. And the vapor cloud could be ignited completely by the venting flame after entering the vapor cloud film. The area above the oblique side of the vapor cloud film firstly ruptured due to the thermal action of the flame, and the flame was released from the ruptured area. Besides, the flame structure experienced a change from blurry to clear as the flame propagated ceaselessly. Accompanied by the gradual disappearance of the irregular flame structure, the flame structure became uniform and dense in the later stage of flame propagation.

Secondly, the temperature of the explosion flow field presented a trend of rising firstly and then decreasing with the increased distance from the venting port after the vapor cloud was detonated by the venting flame. The highest temperature appeared in the central region of the vapor cloud. As for the chain reaction induced by the vapor cloud explosion, the temperature variation also presented a similar trend. As for the ignition condition, the venting flame came in full contact with the vapor cloud and provided the required energy for the explosion.

Thirdly, the explosion pressure increased first and then decreased with the increase in the distance from the venting port. The highest explosion intensity appeared in the central region of the vapor cloud. The P_{max} of the primary and secondary vapor cloud explosions were 0.86 MPa and 0.89 MPa, respectively. However, the moment in which the pressure began to rise presented a trend of decreasing firstly and then rising. The vapor cloud explosion chain reaction induced by the venting flame could produce the high explosion intensity and could not be ignored.

In practice, to effectively avoid and reduce the harm caused by the vapor cloud explosion and its chain reaction induced by the venting flame, three suggestions are proposed based on the above works: (1) In the process of petrochemical equipment layout, the safe distance between devices should be considered to avoid a chain reaction of vapor cloud explosion after gas leakage. (2) The possibilities of the vapor cloud formation and the ignition source emergence outside the container should be reduced as far as possible. (3) Ensure that it should be as far away as possible from the central region of the vapor cloud when the vapor cloud is detonated by an explosion venting flame.

References

1. Cao XY, Ren JJ, Bi MS, Zhou YH, Li YM. Experimental research on the characteristics of methane/air explosion affected by ultrafine water mist. J. Hazard. Mater. 2017; 324(Part B): 489–497.
2. Zhang J. Effect of sodium on gas combustion flame color. Educ. Chem. 2001; 5: 43–44.
3. Zhang Q, Li D. Comparison of the explosion characteristics of hydrogen, propane, and methane clouds at the stoichiometric concentrations. Int. J. Hydrogen Energy 2017; 42(21): 14794–14808.
4. Wang C, Huang F, Addai EK, Dong X. Effect of concentration and obstacles on flame velocity and overpressure of methane-air mixture. J. Loss Prev. Process Ind. 2016; 43: 302–310.

Chapter 9
Experimental Study on the Domino Effect in Explosions Caused by Vertically Distributed Methane Air Vapor Clouds

9.1 Experimental Apparatus and Procedures

9.1.1 Results and Discussions

Figure 9.1 shows the schematic diagram (a) of the experimental setup and the adjustable platform (b) for the vapor cloud experiments. Hemispherical polyvinyl chloride (PVC) films with a thickness of 0.05 mm and diameters of 0.2, 0.4, 0.6, 0.8 and 1.0 m were used to enclose the vapor cloud since a vapor cloud in the shape of a hemisphere has the highest explosion intensity [1]. The experimental procedure consists of three main steps, being the preparation of the hemispherical vapor cloud, adjustment of measuring devices and ignition of the vapor cloud with an ignitor. In the experiments with a single vapor cloud, the hemispherical PVC film was fixed on the center of the bottom aluminum alloy wall. In the experiments with two vapor clouds, the hemispherical PVC films were fixed symmetrically on the aluminum alloy wall with a specified separation distance. The separation distance was adjustable from 0.6 to 1.4 m with a flexible rule. After the PVC film was sealed and fixed on the backboard, the methane/air mixture (volume concentration of methane ranged from 7 to 12%) was prepared through a 20 L spherical vessel, an air compressor and a gas distribution system, which mainly includes a computer and gas distribution instrument, RCS2000.B (Linyin Technology Ltd., China). The gas mixture was injected into PVC bag which later formed the upper or the lower hemisphere. The injection time was also controlled to maintain atmospheric pressure within the hemispherical bag. Each test is repeated at least three times to ensure repeatability of results.

Two pressure sensors, 211B5 (Kistler, Switzerland) with a measuring range of 0–4 bar (Gauge pressure), were used to record the explosion pressure at a distance of 0.2 m away from the top of the hemispherical dome in each test. The data acquisition instrument used during the experiments was a HiOKI 8860-50 with a measurement range of 5 mV–20 V, resolution of 1/100 of the range and maximum frequency of 10 MHz. An infrared thermal camera (IR), FAST M100k with a spectral response

Z. Wang and X. Cao, *Mechanism of Composite Explosion Inhibition and Venting Protections*, https://doi.org/10.1007/978-981-95-4809-5_9

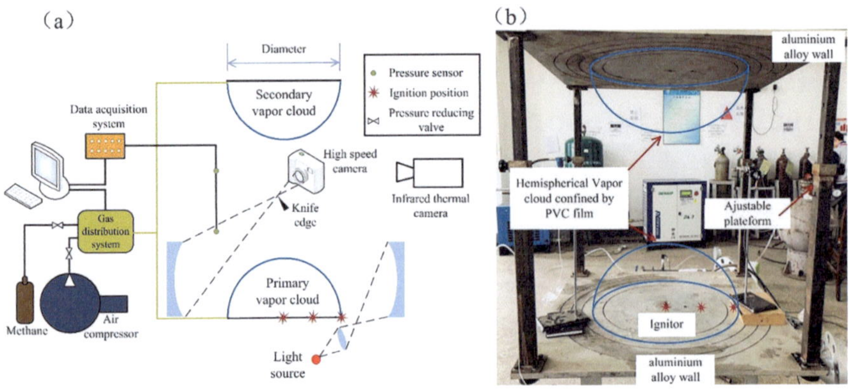

Fig. 9.1 **a** Schematic diagram of the experimental setup and **b** vapor cloud platform

range of 3–4.9 μm, a spatial resolution of 640 × 512 pixels, an exposure time of 10 μs, a frame rate of 490 Hz and resolution of 18 mK, was used in this study to estimate the temperature evolution during the explosion. In addition, a Schlieren system with a scope of 300 mm was employed to record the gas-phase flow field. After the measuring device and data acquisition system were prepared, a high-energy electronic ignitor, XDH-6L (Jiesheng Electromechanical Technology Ltd., China) with ignition energy of 5–30 J was used to ignite the primary bottom vapor cloud. Three different ignition positions, at the center, at the half and right end of the radius, were tried to illustrate the effect of ignition position on the explosion.

Single Hemispherical Vapor Cloud Explosion

Explosion Characteristics

Figure 9.2 shows the evolution of the explosion captured by the IR camera for a single vapor cloud explosion (D = 60 cm) ignited at the center point of the base of the hemisphere. At 10 ms after ignition, the flame front propagated spherically from the ignition point according to visual observations. At 61 ms after ignition, the flame front was semielliptical in shape and the top of the flame broke through the PVC film first. The height of the flame region (vertical direction) was greater than its length (horizontal direction). This outcome mainly results from the effect of the buoyant force induced by the density difference between the hot burned gas and the cold unburnt gas. The upward propagating velocity of the flame front was larger than that in the lateral direction. As a result, the characteristic of the propagating flame changed in the course of the explosion and the flame developed into a semielliptical shape. Due to the asymmetric flame development, there was still unburned gas at the bottom after the flame broke through the PVC film. With the irradiation of the flame and the disturbing action of the pressure wave generated by the explosion,

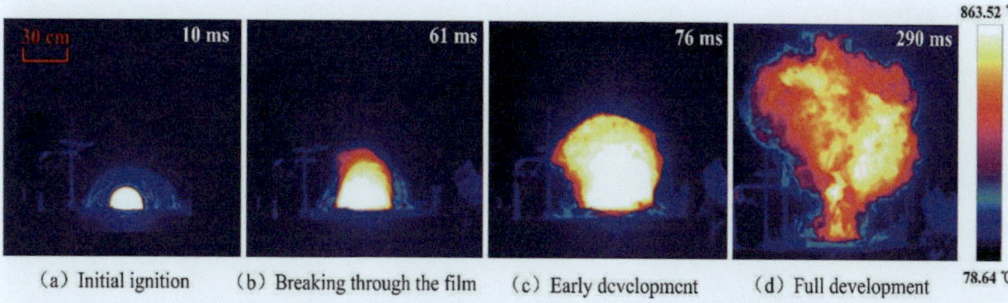

Fig. 9.2 Images captured by IR camera during a single vapor cloud explosion (D = 60 cm) with center ignition

the unburned gas propagated upward, forming a fully developed explosion shown in Fig. 9.2d. In addition, no obvious jet characteristics of flame were found after the PVC film was ruptured. Since the heat needed to melt the film was much less than the heat released during the explosion, the heat loss to the PVC film as well as the impact of the PVC film on the explosion characteristics was assumed to be negligible.

Figure 9.3 illustrates a sequence of Schlieren images for the explosion ignited at the center point. At 5–15 ms, a clear cellular flame and wave front formed by the pressure wave can be observed. As the flame continued to propagate, more cells of smaller size are formed on the flame surface. The flame wrinkling is more pronounced, indicating a higher turbulence intensity. As a result, the burning rate and intensity of the explosion increased with time. At 240 ms, a fully developed cellular flame can be observed. The turbulence reached the strongest level, at which the explosion pressure reaches the maximum value. Thereafter, the cellular flame weakened gradually as the combustible gas burnt out and the flow continued to ascent because of buoyance forces induced from the density difference between the burning region and the surrounding air.

Effect of Methane Volume Concentration

Figure 9.4a shows the pressure histories for explosions (D = 0.6 m) with a methane volume concentration of 7, 9, 10, 11 and 12%. The pressure increases sharply in the initial stage of the gas cloud explosion and decreases gradually after reaching the maximum value. The decreasing rate of the pressure in the attenuation stage is smaller than that reported in the previous studies concerning vapor cloud explosions in the open space [2]. This discrepancy is mainly attributed to the weak explosion and the short distance between the pressure sensor and the vapor cloud. For a vapor cloud explosion without obstacles, the flame propagates at a too slow velocity to produce significant overpressure. Thus, the pressure sensor was fixed at a short distance from the vapor cloud in this study. Unburnt gas participated in the burning process after the vapor cloud explosion was fully developed, leading to a slow decrease in the pressure in the attenuation stage. The maximum pressure and the rupture time of the PVC film versus methane volume concentration are shown in Fig. 9.4b. It

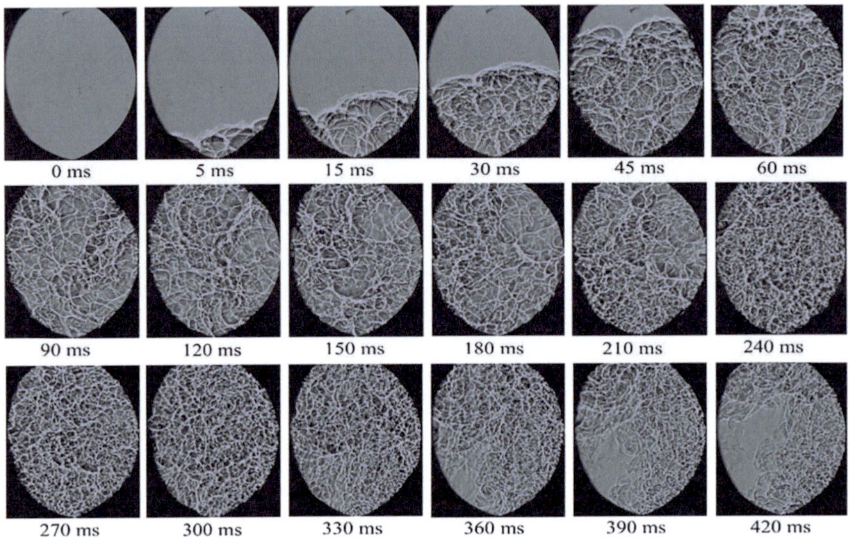

Fig. 9.3 Flame schlieren photographs (ϕ 300 mm) for gas cloud explosion (D = 60 cm) ignited at the center point

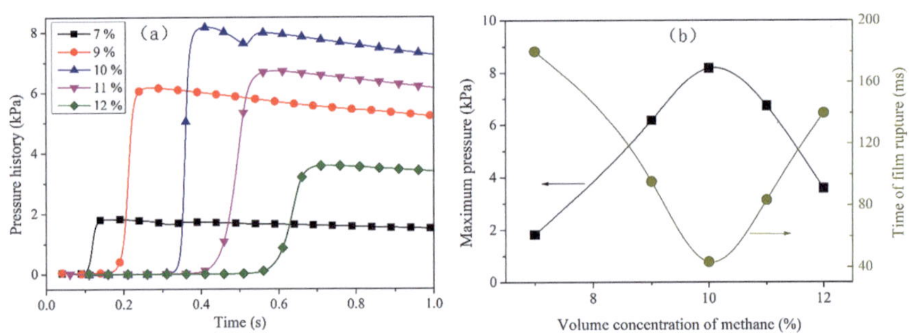

Fig. 9.4 **a** Pressure history versus time for vapor cloud explosions (D = 0.6 m) at different methane concentrations; **b** the maximum explosion pressure and film rupture time versus methane volume concentration

can be seen that the explosion intensity of the vapor cloud increases with volume concentration first and then decreases. The most severe explosion appears at 10%, slightly larger than the stoichiometric value for methane/air, which is consistent with the theoretical explosion pressure calculated by GASEQ and the experimental results in previous studies [3, 4]. At a concentration of 10%, the combustion is complete and the flame temperature is high. Therefore, the flame propagation rate and deflagration intensity reach the maximum value. The time of film rupture shows an opposite trend with maximum pressure, also indicating the severest explosion and the shortest combustion duration at a volume concentration of 10%.

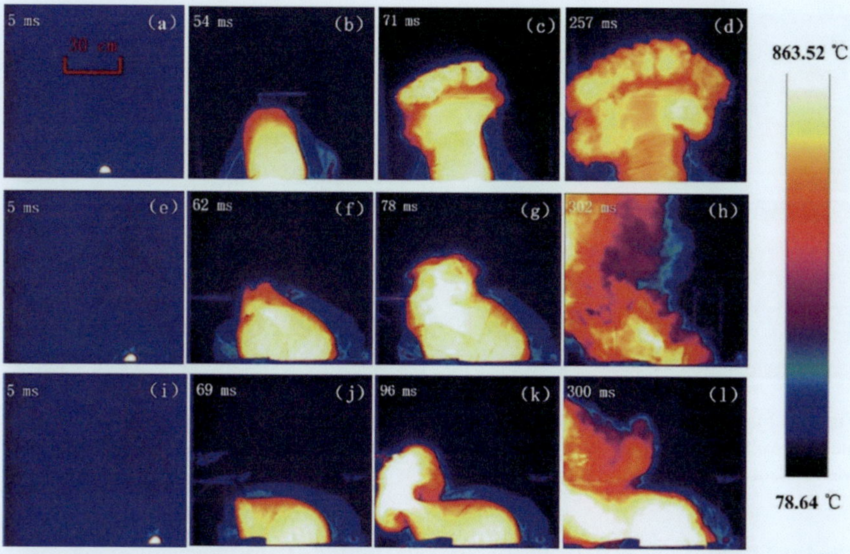

Fig. 9.5 Development process of deflagration flame for vapor cloud explosion ignited at different positions: **a**–**d** at the center; **e**–**h** at half of the radius; **i**–**l** at right end of radius

Effect of Ignition Position and Energy

Figure 9.5 illustrates the development of the flame for the vapor cloud (D = 0.6 m) explosion ignited at the center, at half of the radius and the right end of the radius, respectively. Compared to the vertical development for center ignition, the flame developed differently in the other two cases. The angle between the flame development trajectory and the horizontal direction decreased as the ignitor moved from center to side. Therefore, center ignition was considered as the most possible way to trigger the secondary explosion and was concentrated in the experiments of two vapor clouds. To illustrate the effect of ignition energy, experiments for 10% methane vapor (D = 0.6 m) with center ignition but different energies were conducted. The maximum overpressure hardly changes with the ignition energy considered (5–30 J) in this study, which has been illustrated in the previous study [5]. The influence of ignition energy on the explosion intensity is negligible since the applied energy is three orders of magnitude larger than the minimum ignition energy of a methane/air mixture [6, 7].

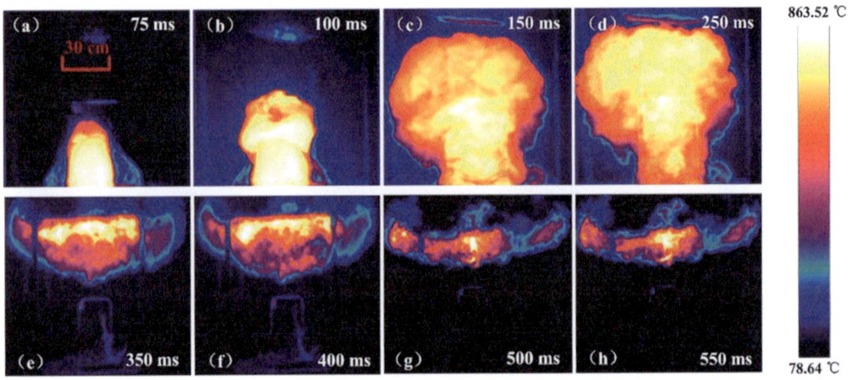

Fig. 9.6 Flame development images for primary and secondary vapor cloud explosions (D = 0.6 m, methane volume concentration of 10%) recorded by IR camera: **a–d** before the second gas cloud was ignited; **e–h** after the second gas cloud was ignited

Chained Explosions of Two Vertically Distributed Hemispherical Vapor Clouds

Characteristics of Chained Explosions

Experiments of two vertically distributed vapor clouds with a methane concentration of 10% and center ignition were conducted, since it refers to the most possible condition to trigger a secondary explosion. Figure 9.6 shows the images of the development processes for primary and secondary explosions recorded with the IR camera. It is obvious that the temperature of the secondary vapor cloud increased gradually during the primary explosion (as shown in Fig. 9.6e–h) before the flame front approached. After the secondary vapor cloud was ignited, the flame at the lower part of the secondary vapor cloud dissipated quickly and propagated mainly around the top aluminum alloy wall, as shown in Fig. 9.6e–h. Chained explosions were considered to be valid if a secondary explosion happened for two of the three repeated tests.

Figure 9.7 shows the temperature histories recorded by the IR camera at five different positions (denoted with red points) around the secondary vapor cloud. It is apparent that there are 2 peaks for each curve, related to the primary and secondary explosions. The temperature at point 3 (the top of the upper hemispherical dome) reached the peak value earliest due to the impact from the flame of the primary explosion. Because of the high temperature and rapid combustion of methane, the flame near point 3 attenuated quickly, leading to an inconspicuous second temperature peak. In contrast, the flame near the ceiling appears to have a slightly lower temperature, but the second temperature rise is larger. In addition, the second peak temperature of points 1, 2 is greater than that of points 5 and 4. This outcome can be inferred from the flame development of the secondary vapor cloud shown in Fig. 9.6e, f. The flame at the left side (corresponding to the side of points 1 and 2) is more intense than that at

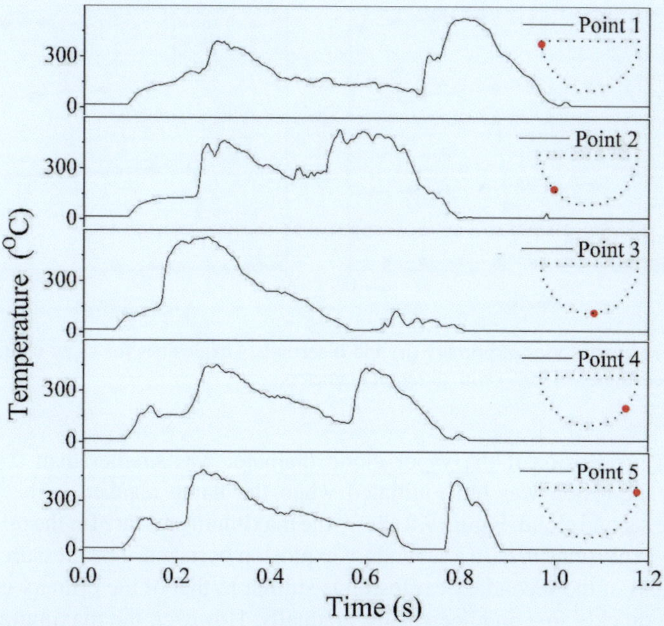

Fig. 9.7 Temperature curves at various measurement points (denoted by the red point) around the secondary vapor cloud (D = 0.6, methane volume concentration of 10%)

the right side (corresponding to the side of points 5 and 4). The asymmetric development of the deflagration flame of the secondary explosion is quite different from the regular gas cloud explosion ignited with a point-source ignitor. This outcome mainly results from the uncertain first ignition point of secondary explosion since the development of the vapor cloud explosion is sensitive to the ignition position. In addition, the fact that the upper gas cloud is unevenly heated due to the disturbing action of the flame front and the pressure wave of the primary explosion would also play a role. The heat sink effect from the aluminum alloy wall reduced the combustion rate of deflagration flame and thus the first peak temperature. However, the second peak temperature near the ceiling was relatively higher, due to the movement of the flame induced by buoyancy. It should be noted that the optical effect due to the absorption and reflection of the PVC film was neglected since the maximum temperature was determined in this study.

Effect of Vapor Cloud Diameter

To investigate the effect of vapor cloud diameter on chain explosions, experiments of two vertically distributed vapor clouds with diameters of 0.2, 0.4, 0.6, 0.8, 1.0 m were executed. The separation distance between the two vapor clouds is set at 1.0 m. According to the experimental results, the secondary explosion was not initiated by

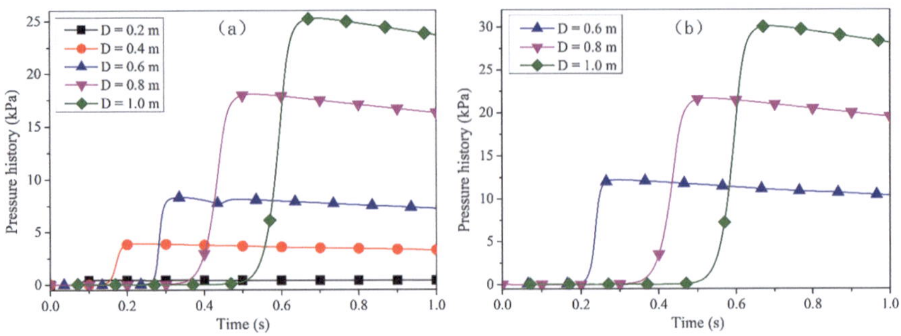

Fig. 9.8 Pressure histories of primary (**a**) and **b** secondary explosions for vapor cloud (methane volume concentration of 10%) with different diameters

the primary explosion if the vapor cloud diameter was smaller than 0.4 m. The secondary explosion was only initiated when the flame reached high enough to contact the second cloud. Figure 9.8 shows the maximum pressure for the primary and secondary explosions in case a secondary explosion occurred. The pressure development process of the secondary explosion is similar to that of the primary explosion, increasing quickly first and decreasing gradually. However, the maximum pressure of the secondary explosion is significantly larger. In addition, it can be seen that no indication of the first explosion is shown in the pressure history of the secondary explosion. This outcome is inferred to mainly result from the weak deflagration and strong ignition of the secondary explosion. The pressure evolution of the secondary explosion is a coupled process, which is initiated by the first explosion and enhanced by the secondary explosion.

Figure 9.9 presents the variation of the maximum explosion pressure with vapor cloud diameter, corresponding to Fig. 9.8. It is seen that the maximum pressures for the primary and secondary explosions increase with diameter. The maximum pressure of the secondary explosion is greater than that of the primary explosion. This is mainly attributed to the increased ignition energy and ignition area for the secondary explosion. The primary explosion was initiated by an electronic spark with relatively little energy. However, for the secondary explosion, the vapor cloud was directly ignited by the deflagration flame of the primary explosion, the ignition energy of which is orders of magnitude larger than the electronic spark. Similar to chemical ignition in a previous investigation, volumetric ignition by the deflagration flame has promoted effects on explosion intensity due to the increased flame front area and burning velocity.

Figure 9.10 shows the Schlieren images of the primary and secondary explosions at the moment when the PVC film was ruptured by the explosion flame. It can be seen that the size of the cellular structure is decreased for the secondary explosion. This discrepancy can be attributed to the increased initial turbulence and pressure because of the primary explosion. Flame generated turbulence by the primary explosion increased the level of initial turbulence within the secondary vapor cloud. With

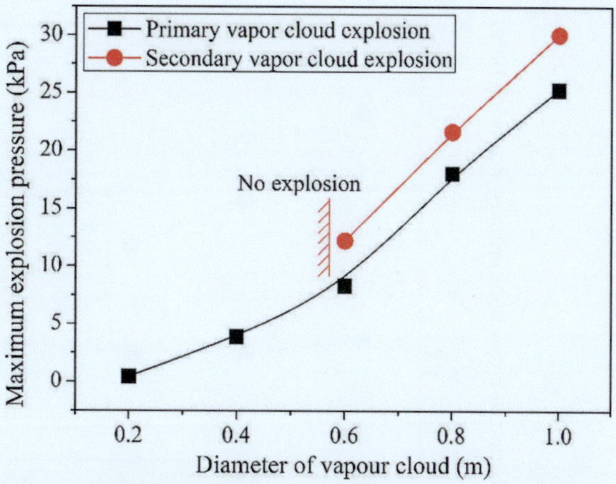

Fig. 9.9 Maximum pressure of primary and secondary explosions versus the diameter of vapor cloud (methane volume concentration of 10%)

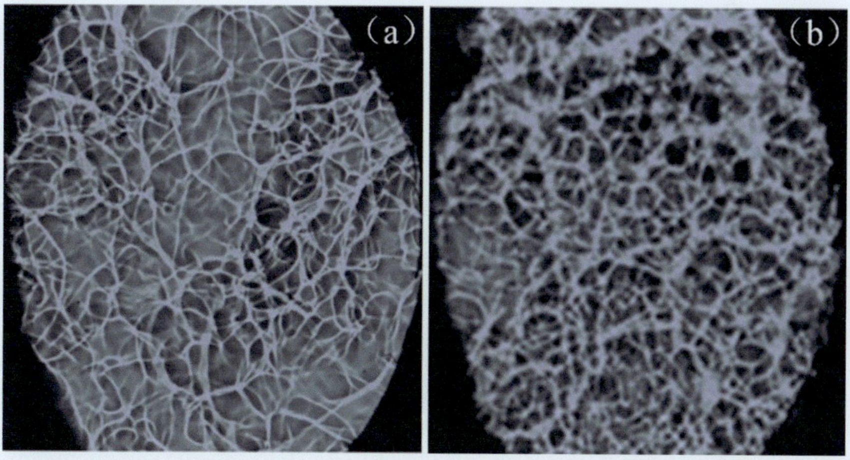

Fig. 9.10 Flame schlieren photographs (ϕ 300 mm) for the primary explosion (**a**) and secondary explosion (**b**)

a disturbed internal flow field, the heat and mass diffusivities increased. The instability of the secondary flame promoted the remaining unburned gas to participate in the reaction and increased flame thickness. As a result, the energy release rate increased, leading to an intensified secondary explosion. The overpressure because of the primary explosion also plays a role since the explosion pressure increases linearly with the increase of initial pressure [8].

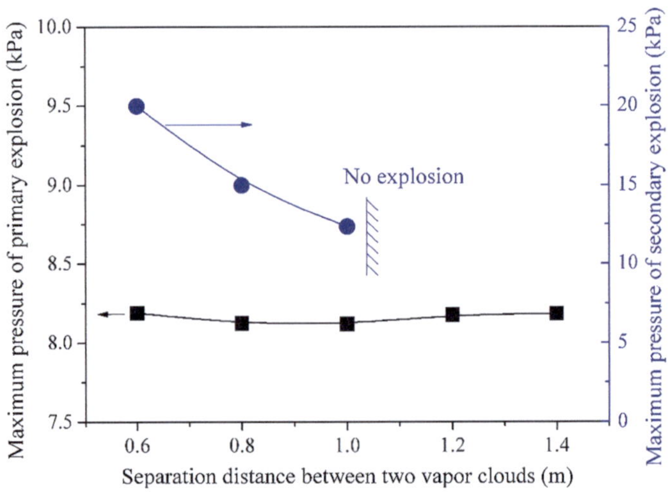

Fig. 9.11 Maximum overpressure for the primary and secondary explosions at different separation distances

Effect of Separation Distance

To investigate the effect of separation distance on the behavior of secondary explosion, experiments for vapor cloud with a methane volume concentration of 10% and a diameter of 0.6 m were conducted at separation distances ranging from 0.6 to 1.4 m. For separation distances larger than 1.0 m, the secondary vapor cloud could not be ignited because the flame height of the primary explosion did not reach high enough. Figure 9.11 shows the maximum pressures for the primary and secondary explosions versus separation distance. It can be seen that the maximum pressure of the secondary explosion decreased with separation distance and is larger than that of the primary explosion. As the separation distance increases from 0.6 to 1 m, the maximum pressure of the secondary explosion decreases gradually from 19.9 to 12.3 kPa and the maximum rate of pressure rise decreases from 475.3 to 383.6 kPa/s. With the increase of the separation distance, the preheating effects and the flame generated turbulence from the primary explosion are decreased, decreasing the flame propagation velocity of the secondary explosion. As a result, the maximum overpressure and the rate of pressure rise decreased with separation distance.

The critical separation distance to trigger a secondary vapor cloud explosion in horizontal chain explosions is 0.3 m for a 0.6 m vapor cloud, smaller than that in this study. In addition, the maximum overpressure of the secondary explosion for a horizontal secondary explosion is smaller than in the case of a vertical secondary explosion for vapor clouds at the same separation distance. Thus, it can be concluded that vertical chain explosions are more severe than the horizontal one at the same separation distance. This outcome can be interpreted from a larger impact of ignition and turbulence induced by the primary explosion flame in the vertical case.

9.2 Conclusions

A series of experiments were conducted on explosions of single and two vertically distributed hemispherical methane vapor clouds in an open space. The effects of ignition position, the diameter of vapor cloud and separation distance on the chain explosions of two vertically distributed vapor clouds were investigated. The conclusion is summarized as follow:

(1) For the primary explosion, the maximum overpressure is independent of the ignition energy considered in this study, which is one order of magnitude higher than the MIE. However, the effect of ignition position on the development of flame is considerable. As the ignitor moves from center to side edge, the angle between the flame development trajectory and the horizontal is decreased.
(2) For the secondary explosion, the maximum pressure and rate of pressure rise of the secondary explosion are larger than that of the primary explosion and increase with vapor cloud diameter, but decrease with separation distance. This phenomenon mainly results from enhanced ignition and the initial turbulence induced by the primary explosion.
(3) Due to the uncertain ignition position of the secondary explosion by the flame of the primary explosion, the development of the secondary explosion is asymmetric, different from the explosion ignited by a point-source ignitor. In general, the secondary explosion triggered by a lower primary explosion is more severe and stronger than that in the horizontal case, therefore, preventions of chain explosions in the vertical case deserve more attention.

References

1. Bjerketvedt D, Bakke JR, Van Wingerden K. Gas explosion handbook. Journal of hazardous materials 1997; 52(1): 1–150.
2. Lv D, Tan W, Liu L, Zhu G, Peng L. Research on maximum explosion overpressure in LNG storage tank areas. Journal of Loss Prevention in the Process Industries 2017; 49: 162–70.
3. Mitu M, Giurcan V, Razus D, Prodan M, Oancea D. Propagation indices of methane-air explosions in closed vessels. Journal of Loss Prevention in the Process Industries 2017; 47: 110–9.
4. Ajrash MJ, Zanganeh J, Moghtaderi B. Influences of the initial ignition energy on methane explosion in a flame deflagration tube. Energy & Fuels 2017; 31(6): 6422–34.
5. Gao W, Zhong S, Miao N, Liu H. Effect of ignition on the explosion behavior of 1-Octadecanol/ air mixtures. Powder technology 2013; 241: 105–14.
6. Ballal D, Lefebvre A. Ignition and flame quenching in flowing gaseous mixtures. Proceedings of the Royal Society of London A Mathematical and Physical Sciences 1977; 357(1689): 163–81.
7. Courtiaud S, Lecysyn N, Damamme G, Poinsot T, Selle L. Analysis of mixing in high-explosive fireballs using small-scale pressurised spheres. Shock Waves 2019; 29(2): 339–53.
8. Askari O, Elia M, Ferrari M, Metghalchi H. Cell formation effects on the burning speeds and flame front area of synthetic gas at high pressures and temperatures. Applied Energy 2017; 189: 568–77.

Correction to: Mechanism of Composite Explosion Inhibition and Venting Protections

Correction to:
Z. Wang and X. Cao, *Mechanism of Composite Explosion Inhibition and Venting Protections*,
https://doi.org/10.1007/978-981-95-4809-5

In the original version of the book, the additional funding has been included in the copyright page. The book has been updated with the changes.

The updated version of this book can be found at
https://doi.org/10.1007/978-981-95-4809-5

Z. Wang and X. Cao, *Mechanism of Composite Explosion Inhibition and Venting Protections*, https://doi.org/10.1007/978-981-95-4809-5_10

MIX
Papier aus verantwortungsvollen Quellen
Paper from responsible sources
FSC® C105338

If you have any concerns about our products, you can contact us on
ProductSafety@springernature.com

In case Publisher is established outside the EU, the EU authorized representative is:
Springer Nature Customer Service Center GmbH
Europaplatz 3, 69115 Heidelberg, Germany

Printed by Libri Plureos GmbH
in Hamburg, Germany